MASS SPECTROMETRY

MASS SPECTROMETRY

Proceedings of the Symposium on Mass Spectrometry

Enfield College of Technology, 5th and 6th July, 1967

Edited by

R. BRYMNER
B.Sc., A.Inst.P.
Faculty of Technology
Enfield College of Technology

J. R. PENNEY
B.Sc., Ph.D., A.R.I.C.
Faculty of Technology
Enfield College of Technology

CHEMICAL PUBLISHING COMPANY, INC.
New York 1969

Mass Spectrometry

ISBN: 978-0-8206-0016-1

Chemical Publishing Company:
www.chemical-publishing.com
www.chemicalpublishing.net

Suggested U.D.C. Number: 543.51
Suggested additional numbers: 535.336.2, 535.853

First American Edition:

Chemical Publishing Company, Inc. - New York 1969

Second Impression:

Chemical Publishing Company, Inc. - 2011

Printed in the United States of America

PREFACE

Our interest in the rapidly expanding field of mass spectrometry prompted us, in the summer of 1966, to consider the organization of a two-day symposium on the subject to be held at Enfield College of Technology the following year.

Preliminary enquiries of many workers in the field met with an immediate response, the interest shown being most encouraging. Our initial decision to cover many aspects of the subject was well received and within a short time most of our invitations to present papers at the symposium had been accepted. We were thus able to formulate a programme at an early stage and receive short contributions within the framework of the sessional topics.

We would like to record our appreciation to all those who presented papers and provided the symposium with such a wealth of personal experience and up-to-date records of achievement in both the design of mass spectrometers and their use in the physical and chemical fields.

We were fortunate indeed that the manufacturers of mass spectrometers and associated equipment responded to our invitations to mount an exhibition concurrently with the symposium and we record our thanks to all those who co-operated.

Mr. G. A. Roberts, Chairman of the Governing Body of the College, presided over a symposium dinner held on the evening of the first day. The guest speaker, Professor Allan MacColl, reviewing the progress of mass spectrometry, flavoured his talk with many personal reflections and anecdotes, thereby ensuring the success of the evening.

The task of the organizers was lightened by the efforts of many willing helpers and we would like to acknowledge the work of the chairmen of sessions, Mr. L. Holland and Professor V. S. Griffiths, and of Professor J. C. Robb for so ably summing up the conference proceedings. We also record our thanks to the Board of Governors of the College for permission to hold the symposium; to Dr. G. S. Brosan, the Principal, for his opening speech of welcome to the delegates and for his encouragement during the preparatory stages; and to our many colleagues for their assistance during the symposium. Last, but by no means least, a very special word of appreciation to Mrs. Renee Hopkins of the secretarial staff, without whom it would all have been very difficult.

PREFACE

In this book, apart from the symbol u for mass units, established symbols have been used throughout. To convert them to the recently introduced S.I. units the following are pertinent:

$$1 \text{ torr} = \left(\frac{101325}{760}\right) \text{Nm}^{-2}$$

$$1 \text{ gauss} = 10^{-4}\text{T}$$

$$1 \text{ tesla} = 1\text{Wm}^{-2}$$

Enfield College of Technology

R. Brymner
J. R. Penney

CONTENTS

SESSION 4

CHEMICAL APPLICATIONS II

Chairman, J. R. Penney, Enfield College of Technology

SESSION 1
DESIGN ASPECTS

1.1

SINGLE FOCUSING MAGNETIC DEFLECTION MASS SPECTROMETERS

R. M. Elliott

1.1.1. INTRODUCTION

MASS spectrometers employing deflection in a magnetic field are widely used because of their reproducibility, stability and ease of adjustment for optimum performance. Ions formed in the ion source are accelerated into a beam and enter the magnetic field with essentially common kinetic energies and directions although the spread in both energy and direction is nevertheless significant. Ions of different masses are deflected along circular paths of different radii and the separate ion beams can be collected and measured in turn. By proper choice of the shape of the magnetic field boundaries, ions emerging from the ion source in different directions are brought to a focus at the collector (single, or direction focusing).

Using a magnetic field alone, the defocusing effects of the spread in energy cannot be eliminated and focusing for direction and energy (double focusing) generally requires the use of an additional electrostatic deflecting field.

In single focusing instruments the irreducible energy spread of some tenths of an electronvolt effectively limits the attainable resolution to values of the order of 10^4.

There are many general descriptions of magnetic deflection mass spectrometers in the literature[1–7], dealing with all aspects, including ion sources, collectors, vacuum systems, electronic supplies, etc., and no attempt is made to cover this ground again here. Instead, this paper concentrates on one aspect of performance, the attainable resolution, and considers the influence of various instrumental parameters in limiting resolution. Two specific instruments are described, one using the 180° deflection principle and the other using a 90° magnetic sector field.

The geometries of these systems are shown in *Figures 1.1.1* and *1.1.2* respectively. The 180° configuration has the following features:

(*i*) The source and collector are both immersed in the magnetic field, making for excellent stability and reproducibility but restricting accessibility to some extent.

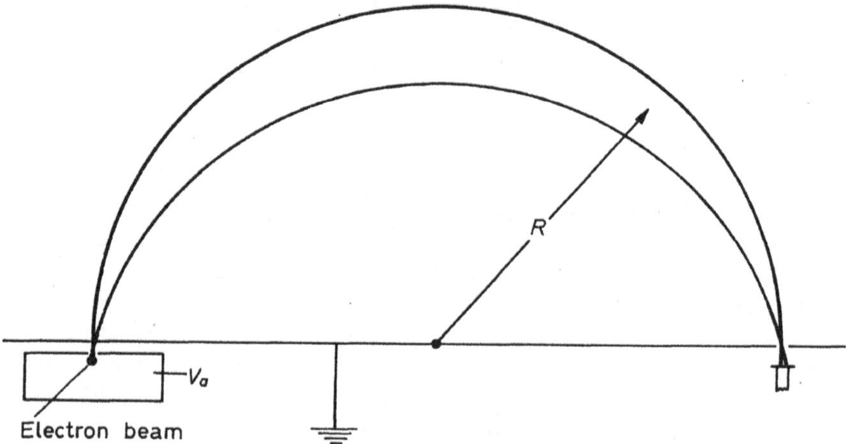

Figure 1.1.1. Focusing in a 180° *magnetic field*

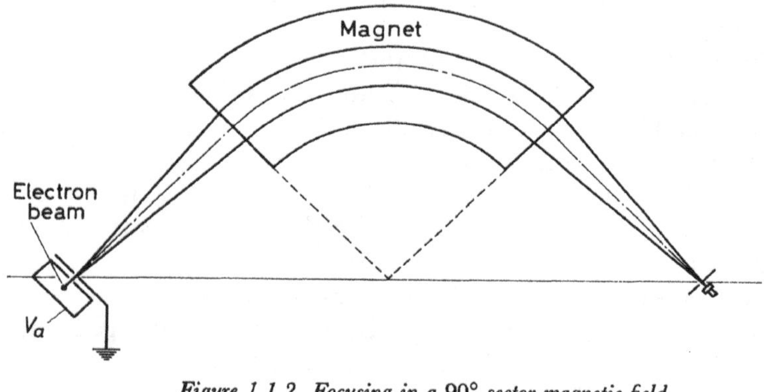

Figure 1.1.2. Focusing in a 90° *sector magnetic field*

(*ii*) The analyser system is very compact, but the magnet pole-face area is relatively large compared with the ion beam radius.

(*iii*) The path length is very short, giving the ions little chance of being lost by drifting in the direction of the magnetic field. In consequence the transmission characteristics are good and mass discrimination is small.

The 90° sector shown in *Figure 1.1.2* is a common choice and 60° is another sector angle frequently used. Sector instruments have the following features:

(*i*) The source and collector are remote from the magnetic field, giving good accessibility.

(*ii*) The magnet pole-face area is relatively small compared with the ion beam radius and, as the sector angle decreases, so does the magnet size. However, as the sector angle decreases the overall size of the analyser increases, giving a longer ion path and consequently more serious mass discrimination under voltage scanning conditions.

1.1.2. RESOLUTION

In considering the ultimate resolution to be expected from a single focusing instrument of a given size, many of the factors involved can be calculated in a straightforward fashion, but there are always some which are impossible to calculate or to predict. For example the effects of electrostatic charges on insulating layers, which build up in time on surfaces in the ion source or analyser due to ion or electron bombardment of adsorbed organic material, are ignored in the calculations which follow, though in practice they can be as large as any other single factor.

A general expression for the resolution of a single focusing instrument with deflection in a uniform magnetic field can be written as follows:

$$\text{Resolution} = \frac{r_m}{W_c + M_m W_s + f\left(r_m \dfrac{\Delta V}{V}, \alpha^2 r_m, \dfrac{z^2}{r_m}\right)} \quad \ldots. (1.1.1)$$

where r_m is the radius of the ion path in the magnetic analyser, W_c is the width of the collector (image) slit, M_m is the linear magnification in the magnetic analyser, W_s is the width of ion source (object) slit, ΔV is the spread in energy of the ion beam emerging from the source, V is the applied accelerating voltage, 2α is the angle of divergence of the ion beam and $2z$ is the linear extension of the ion beam in the direction of the magnetic field. It is simplest to assume:

(*i*) A symmetrical arrangement as in *Figures 1.1.1* and *1.1.2* since most practical arrangements are symmetrical, or nearly so, and the effect of asymmetry on the resolution is small in any case. For symmetrical systems $M_m = 1$.

(*ii*) That the function f is the arithmetical sum Σ. This implies that the contributions of the various aberrations to the width of the refocused ion beam add together arithmetically. This is not strictly true but is sufficiently accurate for this treatment, since only general conclusions will be drawn.

The expression then becomes:

$$\text{Resolution} = \frac{r_m}{W_c + W_s + r_m \dfrac{\Delta V}{V} + \alpha^2 r_m + k \cdot \dfrac{z^2}{r_m}} \quad \ldots . (1.1.2)$$

$$\frac{1}{\text{Resolution}} = \frac{W_c}{r_m} + \frac{W_s}{r_m} + \frac{\Delta V}{V} + \alpha^2 + k \cdot \frac{z^2}{r_m^2} \quad \ldots . (1.1.3)$$

where $k = 1$ for a symmetrical sector instrument and $k = 0$ for a 180° instrument.

It is convenient to consider the reciprocal of the resolution, as in equation 1.1.3, since each factor appears as part of a simple addition sum and its significance can be readily seen. Each factor is conveniently calculated in parts per million (p.p.m.), the summation carried out in those units and the reciprocal then taken to give the final resolution figure.

Table 1.1.1. Instrumental Parameters

Configuration	*Inches*				*Radians*	*Volts*	
	r_m	W_s	W_c	$2z$	α	V	ΔV
No. 1	12	⩾0·0001	⩾0·0001	0·05	0·005	⩽10,000	0·2
No. 2	12	⩾0·0001	⩾0·0001	0·2	0·010	⩽10,000	0·4
No. 3	2	0·001	0·001	0·4	0·014	500	0·4

In order to assess the meaning of this expression in practical terms, three hypothetical analyser configurations are considered. They are, as numbered in *Table 1.1.1*:

No. 1. An idealized instrument of radius 12 in. The size is probably approaching the maximum desirable for the typical laboratory room. The slits (W_c, W_s) are very narrow and all the angular and linear dimensions (α, z) of the ion beam are severely restricted in order to achieve maximum resolution. The energy spread (ΔV) of 0·2 eV corresponds to a carefully cleaned ion source, free from serious contamination and operated within a restricted range of ion source parameters such as ion repeller voltage.

This concentration on resolution alone results in drastically reduced sensitivity.

No. 2. An example of a practical compromise between resolution on the one hand and sensitivity and ease of operation on the other. Many other combinations of parameters could be chosen but, in general, an increase in resolution must be balanced against a consequent loss in sensitivity.

No. 3. A small, simple 180° instrument. The overall dimension of the ion path is eight times less than for Nos. 1 and 2 and wider slits and low voltages make for simplicity in setting up and operation.

Table 1.1.2. Contributions to Aberrations

Configuration	*Contribution* (p.p.m.)						
	W_s/r_m	W_c/r_m	$k.z^2/r_m^2$	α^2	$\Delta V/V$	$\frac{10^6}{Resolution}$	*Resolution*
No. 1	⩾8	⩾8	4	25	⩾20	⩾65	⩽15,000
No. 2	⩾8	⩾8	64	100	⩾40	⩾220	⩽4,500
No. 3	500	500	—	200	800	2000	500

Tables 1.1.1 and *1.1.2* give only a very limited picture of the possible combinations available but show that resolution of several hundred should be practical with quite simple instruments, while resolution up to several thousand is practical while retaining adequate sensitivity. Resolution well above 10,000 is possible, in principle, but probably at the expense of sensitivity and in this region a double focusing instrument is more likely to be a practical solution.

1.1.3. A 180° MASS SPECTROMETER—THE AEI MS10

An example of a 180° deflection instrument of simple construction and small dimensions is the AEI MS10. This is a 2 in. radius instrument, constructed as shown in *Figure 1.1.3.*

The ion source is immersed in the magnetic field and the electron beam is therefore well collimated. The electron beam current can be varied between 10 μA and 300 μA and the electron energy between 5 V and 105 V.

The ion accelerating voltage can be varied between 2000 V and 20 V. With the 1850 G magnet used in the simpler forms of the instrument, this gives a mass range of 2–200 and, since the resolution in this form, with a 0·005 in collector slit, is 115 on a 10% valley definition, useful mass spectra can be obtained up to mass 200. Despite the low ion accelerating voltage at the high mass end, mass discrimination is not serious because of the high field in the ion source and the short ion path.

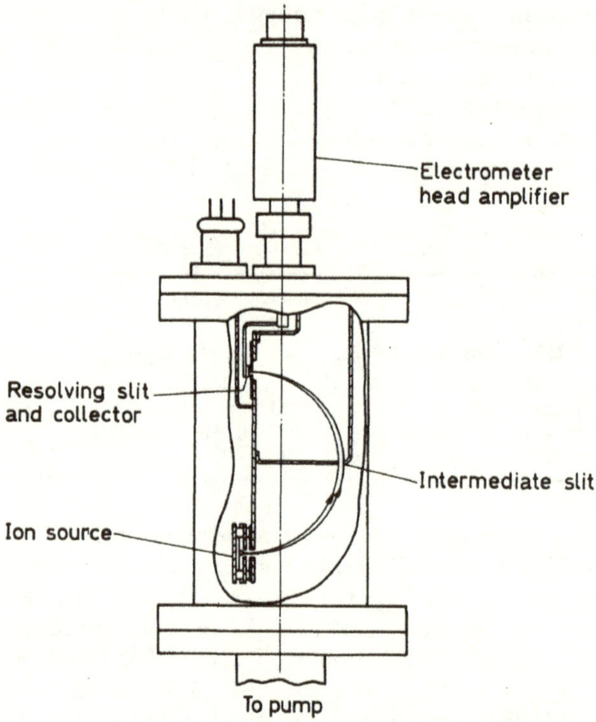

Figure 1.1.3. Schematic of the analyser, AEI MS10 mass spectrometer

Much better performance is obtained, particularly at high mass, using either a larger permanent magnet (4000 G) or an electromagnet giving up to 9000 G. With the electromagnet the mass range is extended to 800, the 10% valley resolution is greater than 350 (see *Figure 1.1.4*) and peaks at adjacent mass numbers can be distinguished up to mass 600.

The controls for this instrument are particularly simple. In its permanent magnet form all the controls required are contained in a single unit and the mass number is selected merely by tuning the mass dial to the appropriate number. The mass dial can also be motor driven giving an automatic scan of 50 s, 3½ min or 13 min, depending on the motor fitted. For operation with the electromagnet, an additional control unit provides electronically controlled scans over a wide range of scan speeds.

In its standard form the instrument has been used for a very wide variety of applications including fundamental studies of the kinetics of chemical reactions, ionization potentials, isotopic abundance measurements in geology[8], biological tracer studies and all kinds of gas analysis applications[9, 10, 11].

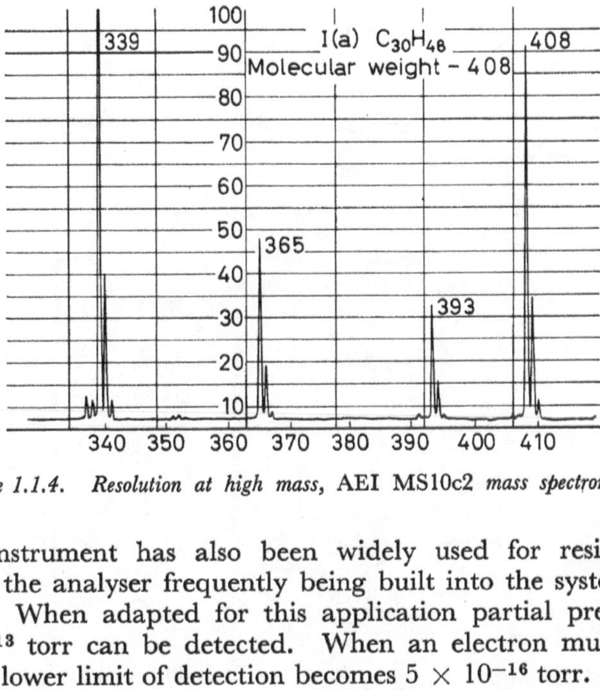

Figure 1.1.4. Resolution at high mass, AEI MS10c2 *mass spectrometer*

The instrument has also been widely used for residual gas analysis, the analyser frequently being built into the system being studied. When adapted for this application partial pressures of 5×10^{-13} torr can be detected. When an electron multiplier is used the lower limit of detection becomes 5×10^{-16} torr.

In high resolution form the instrument is capable of identifying organic compounds of molecular weight up to 600. A reservoir inlet system is available for the more volatile compounds. For involatile materials such as steroids the sample is introduced into a tube connected through a valve into the analyser tube and evaporated directly into the analyser without an intervening leak. The MS10 has also been used to identify gas chromatograph eluents[12, 13] in conjunction with an interrupted elution chromatograph and an infra-red spectrophotometer.

The MS10, therefore, is a simple and compact instrument with a wide variety of applications. In practice it achieves a resolution of 350 or greater, approaching quite closely the theoretical value calculated in *Table 1.1.2.*

1.1.4. A 90° SECTOR MASS SPECTROMETER—THE AEI MS12

The AEI MS12 is a single focusing instrument with a 90° magnetic sector and an ion path radius of 12 in. A schematic diagram of the tube is shown in *Figure 1.1.5.*

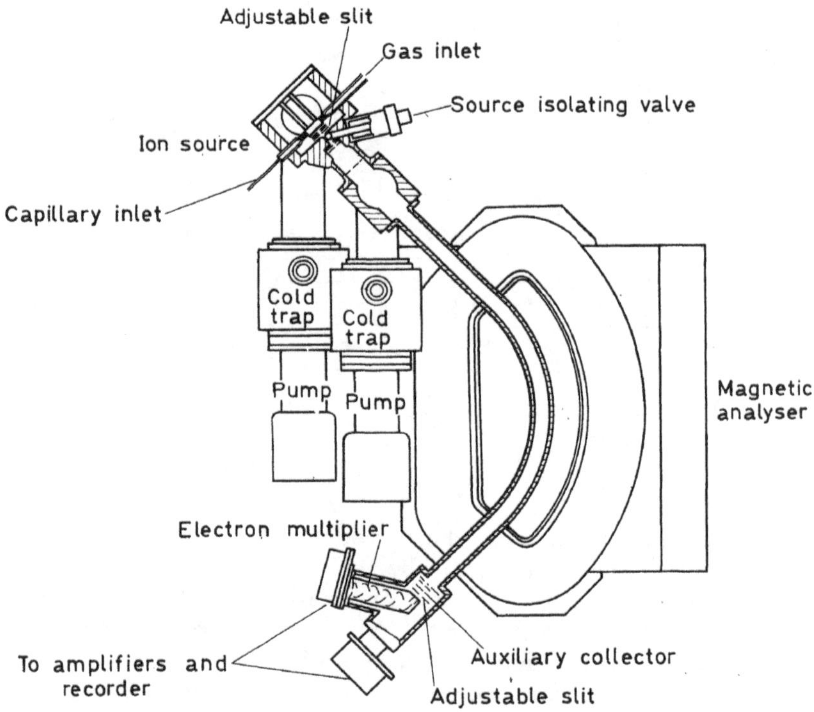

Figure 1.1.5. Schematic of the analyser, AEI MS12 *mass spectrometer*

The ion source region has four inlet ports which are directly in line with the ionization region and can be used for connecting up a variety of inlet systems including a direct sample insertion lock, reservoir inlet systems and chromatographic inlets such as flow splitters or Biemann separators.

The ion source can be completely isolated from the rest of the vacuum system for venting to atmosphere. The ion source and analyser are differentially pumped and the instrument can, therefore, be operated with relatively high sample pressures (of the order of 10^{-4} torr).

Continuously variable source and collector slits allow the resolution to be easily controlled from outside the vacuum. In each case the minimum slit width is 0·00005 in. A total ion current monitor is fitted. This intercepts about 20% of the ion beam and can be used for monitoring sample flow, especially where the instrument is directly linked to a gas chromatograph when it provides a chromatograph record.

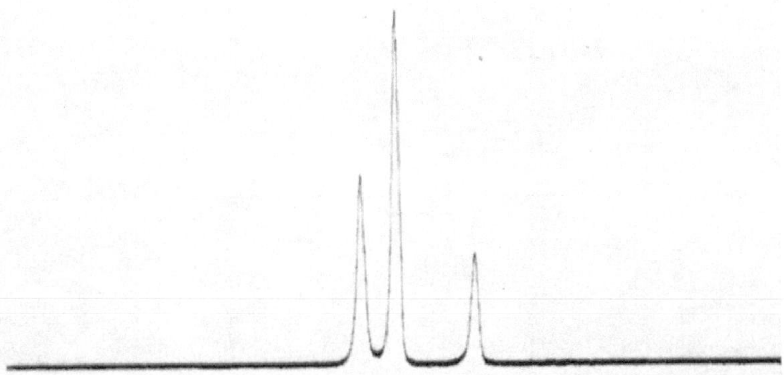

Figure 1.1.6. Triplet at mass 28 obtained on the AEI MS12 *mass spectrometer; from left to right the peaks are due to* CO, *and* N_2, C_2H_4; *the mass difference between* CO *and* N_2 *is one in 2540 and the resolution shown is over 4000*

The ion beams are normally detected by an eleven-stage electron multiplier in order to attain the highest sensitivity and speed of response. The bandwidth of the recording system is 10 kHz using a tape recorder, but when the ultra-violet recorder is used, the galvanometers impose a limit of about 4 kHz.

The mass spectrum is scanned by sweeping either the magnetic field or the accelerating voltage. The fastest magnetic scan is 2 s per mass decade and the fastest voltage scan is 0·5 s per mass decade. In each case the scan speed is adequate for use of the instrument directly linked to a chromatograph. For analytical

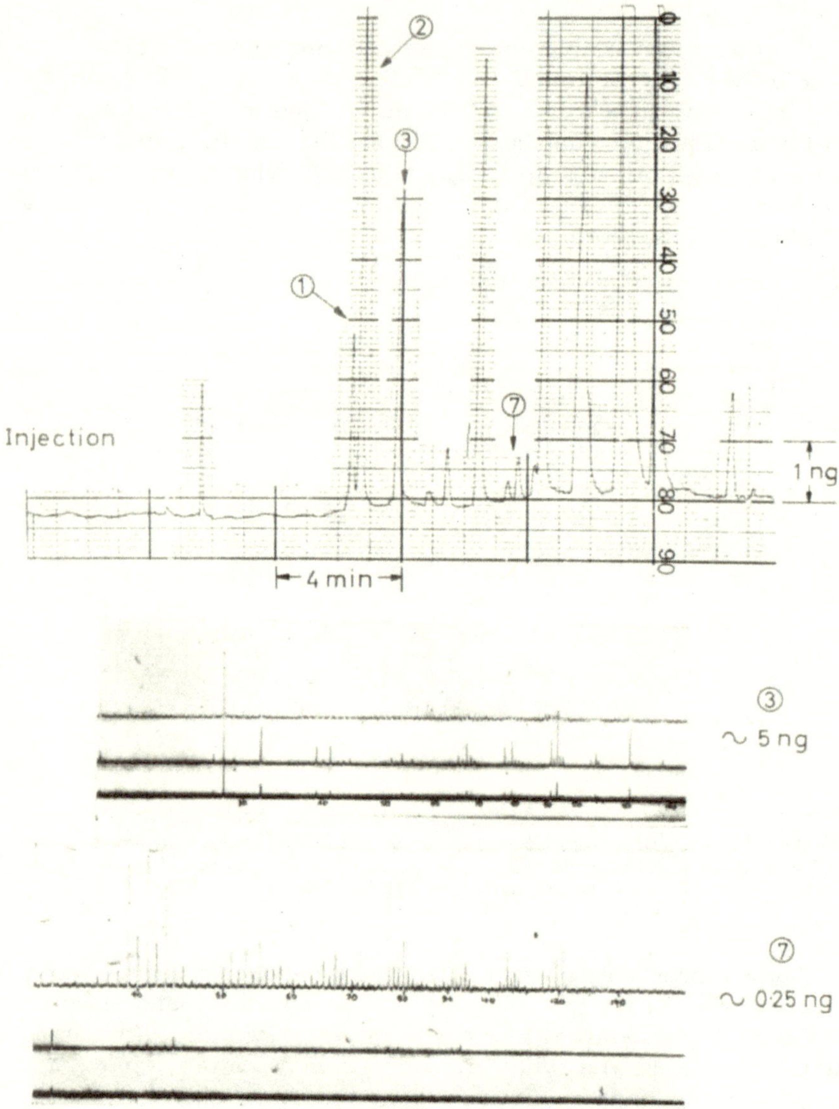

Figure 1.1.7. Part chromatogram of 3 µl *of red thyme oil, and mass spectra of components 3 and 7; column split ratio 1:300, i.e.* 0·01 µl *on column; capillary column, polypropylene glycol,* 50 m × ½ mm

applications the magnetic scan is generally preferred because of the loss in sensitivity at high mass when voltage scanning is used.

The normal ion accelerating voltage used is 8000 V. At this voltage the mass range is from 1 to 900 and the guaranteed resolution is 7500 on the 10% valley definition. At 2 kV the mass range is from 4 to 3600. A resolution of 12,500 is demonstrated in *Figure 1.1.6.* which shows the spectrum of the doublet CO, N_2, where the mass difference is in 2510.

The principal application of the MS12 is the identification of organic compounds. Identifiable spectra can be obtained from less than 1 ng of an unknown compound of molecular weight 300, uniformly evaporated in the source, during the period of the scan, at a resolution of 1000.

When the unknown compound is a gas chromatograph eluent, spectra suitable for identification can be obtained from less than 3 ng of the component injected onto the column[14].

An example is shown in *Figure 1.1.7*, where the part chromatogram of red thyme oil was obtained on a pen recorder connected to the total ion current monitor. Two of the mass spectra obtained are also shown, the lower spectrum corresponding to only 0·25 ng of component entering the ion source.

The MS12, therefore, is an example of a high performance single focusing mass spectrometer with sufficient sensitivity to detect nanogramme quantities of organic compounds at a resolution of 1000, yet having a resolution capability of well over 5000.

The author wishes to acknowledge the contribution of many of his colleagues in the Scientific Apparatus Department of Associated Electrical Industries Limited to the various theoretical and practical results described in this paper.

REFERENCES

1 Ewald, H. and Hintenberger, H., *Methoden und Anwendungen der Massenspektroskopie*, Verlag Chemie, Weinheim, 1953

2 Inghram, M. G. and Hayden, R. J., 'A Handbook of Mass Spectrometry', *Natn. Acad. Sci.–Natn. Res. Coun., Nucl. Sci. Ser.*, Rep. 14, 1954

3 Duckworth, H. E., *Mass Spectroscopy*, Cambridge University Press, London, 1958

4 Beynon, J. H., *Mass Spectrometry and its Applications to Organic Chemistry*, Elsevier, Amsterdam, 1960

5 McDowell, C. A. (Ed.), *Mass Spectrometry*, McGraw-Hill, New York, 1963

6 Errock, G. A., *Mass Spectrometry*, Reed, R. I. (Ed.), Academic Press, New York, 1967

7 Brunnee, C. and Voshage, H., *Massenspektrometrie*, Verlag Karl Thiemig, Munich, 1967

8 Farrar, E., MacIntyre, R. M. and York, D. *Am. Soc. Test. Mater. 12th A. Conf. Mass Spectrom. Allied Topics*, Montreal (1964).

9 Hazelby, D. *J. sci. Instrum.* 42 (1965) 486

[10] ASPINAL, M. L. *Analyst, Lond.* 91 (1966) 1078
[11] CAMPAU, R. M. and NEERMAN, J. C. *S.A.E. Automotive Engng Congr.*, Detroit (1966)
[12] SCOTT, R. P. W., FOWLIS, I. A., WELTI, D. and WILKINS, T., *Proc. int. Symp. Gas Chromat.*, Butterworths, London, 1966
[13] ELLIOTT, R. M. and RICHARDSON, W. J. *Am. Soc. Test. Mater. 15th A. Conf. Mass Spectrom. Allied Topics*, Denver (1967)
[14] CHAPMAN, J. R. *Am. Soc. Test. Mater. 15th A. Conf. Mass Spectrom. Allied Topics*, Denver (1967)

DISCUSSION

J. F. J. TODD (UNIVERSITY OF KENT)
What is the detectable limit in terms of parts per million on the MS12 so that it can be compared with the MS10?

R. M. ELLIOTT, *in reply.* The MS12 is not designed primarily for his application and the sensitivity is usually defined in other ways. The basic sensitivity is lower than the MS10, but with the electron multiplier the overall sensitivity is very high. In theory the MS12 should be able to detect less than 1 part in 10^8 of one gas in another. However, in practical cases, the limit is more often than not set by overlap between the spectra of the impurity and the major component, and each individual case has to be considered separately.

H. W. WERNER (PHILIPS, EINDHOVEN)
What is the bandwidth of the MS10 at an input resistance of $10^{12}\Omega$ to obtain a peak-to-peak noise of 4×10^{-16} A? Am I correct in guessing from the records you have shown, that it was a few hertz?

R. M. ELLIOTT, *in reply.* The bandwidth of the MS10 amplifier is determined by the capacity of a few picofarads across the input resistor, so that with $10^{12}\Omega$ the bandwidth is well below 1 Hz.

In order to gain the advantage of the low input current noise when using the high resistance, a wide collector slit is used, so that when the mass dial is set to a particular mass value the peak top is found with absolute certainty.

1.2

THE DESIGN OF DOUBLE FOCUSING MAGNETIC DEFLECTION INSTRUMENTS

G. A. Errock

1.2.1. INTRODUCTION

THIS paper considers only the design of instruments employing magnetic deflection for the separation of ions according to the ratio of their mass to the charge which they carry. Other types of instrument exist and their design and applications have been described by Brunnee[1] and Scheidegger (see p. 37).

Instruments employing only a magnetic deflection field after the electrostatic accelerating system incorporated in the ion source, are normally chosen to give first order angular focusing for the ions reaching the collector, in addition to the mass dispersion. Instruments of this type have been described by Elliott (see p. 3).

From the basic mass spectrometer formula:

$$\frac{mv^2}{R} = Hev$$

where m is the mass of the ion, v is its velocity, R is the radius of curvature of the ion path in the magnetic field H and e is the charge carried by the ion, one derives $mv = HeR$, i.e. the magnetic field sector acts as a momentum separator. Consequently the ability of the system to separate ions of different masses, in other words its resolution, depends on the energy spread of the ions leaving the source. Such instruments are said to be 'single focusing' because they possess a first order angular focusing property but do not provide velocity focusing. It has been shown by Elliott (see p. 6) that resolutions up to 10,000 or so can be attained with a single focusing instrument, using an ion accelerating voltage of 8 kV and a radius of curvature of 12 in. in the magnetic field, with an electron bombardment ion source.

A need can arise for greater resolution when the mass separation of ions in the spectrum is less than 1 in 10,000. In addition some forms of ion source, for example the vacuum spark source and the gas discharge source, give ions of very wide energy spread, of the order of several hundred volts, hence the resolution obtained with a single focusing instrument would be too small even to separate adjacent masses in the low mass region.

1.2.2. DOUBLE FOCUSING

A combination of a magnetic sector with one or more electrostatic field sectors can be arranged to give first order velocity focusing and angular focusing simultaneously; such arrangements are said to be 'first order double focusing'.

The form of electrostatic field usually employed is the radial field between two coaxial sector-shaped cylinders, as shown in *Figure 1.2.1.* The mean radius of curvature of the two cylinders is

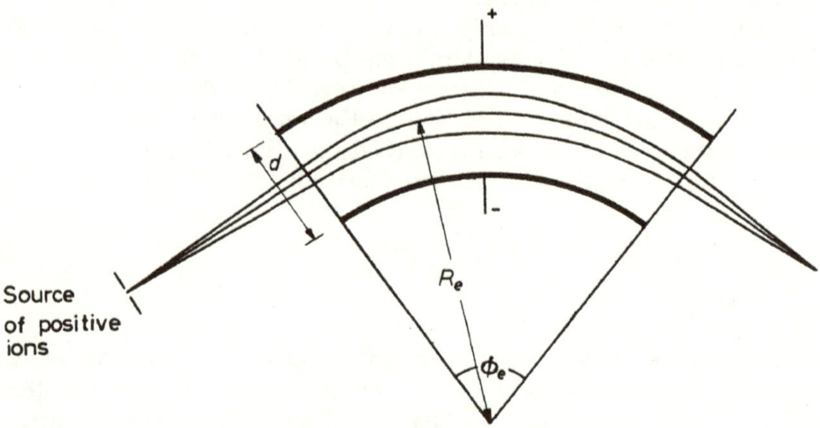

Figure 1.2.1. Electrostatic analyser

R_e. The outer cylinder is connected to a positive potential with respect to earth and the inner cylinder to an equal negative potential. A radial electrostatic field is set up between the two cylinders, the potential along the central path being zero.

An ion entering the electric field at right angles to the boundary, travels in a circular orbit throughout the field if its energy is such that the centrifugal force acting on it is balanced at all times by the radial electrostatic force. The equation of motion of the ion becomes:

$$\frac{mv^2}{R_e} = eE_0$$

where E_0 is the electrostatic field acting at the mean radius.

The radius of curvature of the path of the ion is seen to be dependent on its energy, but independent of its mass. An ion having a slightly greater energy, also entering the field along the centre line, will travel in a path of greater radius and will emerge from the sector at a position above the centre line. The action is analogous to that of a prism and ions having different energies, which enter the field along the centre line, emerge from the sector at positions which depend upon their energies. A lens effect also occurs along with the prism effect so that ions of a given energy, which leave a point source and enter a sector field, can be brought to a focus to form a point image, providing that their angular divergence, with respect to the centre line, is sufficiently small.

A general theory of electrostatic field sectors, derived by Herzog[2], shows that a given focal length and magnification of an electrostatic lens can be produced by suitable choice of geometry. Ions from a line source, used as the object for such a sector, give rise to a series of line images in the focal plane and form an energy spectrum of the source. A narrow slit, placed in the focal plane, can be used to transmit ions in a narrow band of energies centred on the mean energy of the source. If such an energy filter were used as the ion source for a magnetic sector instrument then its resolution would be improved. This, however, would be accompanied by a loss in sensitivity, since ions outside the selected narrow energy band would not reach the magnetic analyser.

Fortunately it is possible to avoid this loss in sensitivity by choosing such a configuration of electrostatic and magnetic fields that the energy dispersion, suffered by the ions in travelling through the electrostatic sector, is annulled by the energy dispersion suffered in the magnetic sector. It is also possible to arrange that angular focusing occurs in the combination as well as energy focusing, so that the expression describing the image position contains no first order terms involving the angle of divergence (α) or the energy spread (β). These so-called 'first order double focusing' systems do, however, contain terms involving α^2 and β^2 and higher orders.

1.2.3. DOUBLE FOCUSING DESIGNS

Dempster[3], Bainbridge[4] and Nier[5] have described designs of instrument in which the first order double focusing conditions are met for one radius of curvature in the magnetic field. Johnson and Nier[6] extended the parameters considered to include second order angular focusing, giving rise to a design, using asymmetrical sectors, which allows the use of a relatively large angle of divergence in the initial beam. This results in an improvement in sensitivity for a given resolution.

The mass spectrum is obtained in these instruments by scanning the ions successively across a fixed collector by sweeping either the magnetic or voltage fields.

A further form of instrument design, originated by Mattauch and Herzog[7], provides first order double focusing for all radii of curvature in the magnetic field and, therefore, for all masses within the range of the instrument, simultaneously. The image plane is flat in this design so that a photoplate can be used as the detector, providing simultaneous integration of all ion beams. The design can also be made to provide second order angular focusing for one particular radius of curvature together with first order double focusing for all masses. This feature can be of value with a Mattauch–Herzog instrument fitted with an electrical detector. In the true Mattauch–Herzog design the source slit is placed at the principal focus of an electrostatic sector of 31° 50′. The image formed by this sector is focused at infinity and, after undergoing a deflection through 90° in the magnetic sector, the ions converge to form a mass dispersed image at the principal focus which coincides with the boundary of the magnetic field.

1.2.4. ORGANIC ANALYSIS REQUIREMENTS

In order to assess the features of a practical instrument the requirements of organic analysis are first considered.

1.2.4.1. Resolution

A need for high resolution exists, even at low masses, as a result of the 'mass defect' which exists between the actual mass of an atom and the nearest whole number mass, which is a function of the binding energy of the nuclei.

When interpreting the low resolution spectrum of a sample material containing, for example, carbon, hydrogen, nitrogen and oxygen atoms, the identity of a peak at a certain nominal mass may be in doubt because a number of different arrangements of atoms can give rise to the same nominal mass. For example the combinations of CO^+, N_2^+, CH_2N^+ and $C_2H_4^+$ all have a mass of very nearly 28 atomic mass units (u). However, when one considers the exact atomic masses on the scale based on ^{12}C, the exact masses are H = 1·0078246, N = 14·0030738 and O = 15·9949141, hence the masses of the ions considered are seen to be CO = 27·994914, N_2 = 28·006147, CH_2N = 28·018723 and C_2H_4 = 28·031299. The smallest fractional mass difference between members of this group of ions is 1/2490 occurring between the 'doublet' N_2^+–CO^+, hence

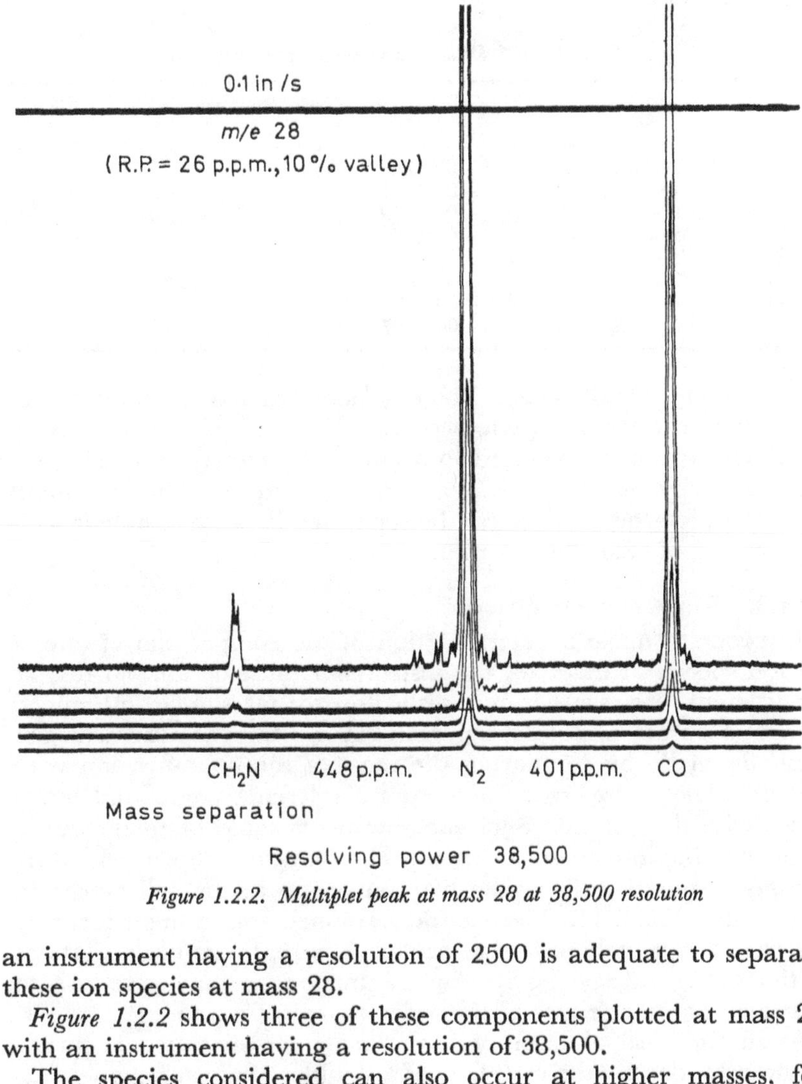

Figure 1.2.2. Multiplet peak at mass 28 at 38,500 resolution

an instrument having a resolution of 2500 is adequate to separate these ion species at mass 28.

Figure 1.2.2 shows three of these components plotted at mass 28 with an instrument having a resolution of 38,500.

The species considered can also occur at higher masses, for example, as $R \cdot CO^+$ and $R \cdot N_2^+$, where R represents any combination of atoms; the resolution required to separate these ions increases linearly with mass.

Table 1.2.1 shows other commonly occurring doublets and also shows the maximum mass at which they can be separated with a resolution of 25,000. If one considers that the majority of samples to be analysed are of molecular weight 600 or less, then the resolution required is of the order of 50,000.

Table 1.2.1. Doublets and Mass Separations

Mass	*Doublet*	*Separation ΔM*	*Maximum mass at which doublet could be identified*
14	$N–CH_2$	−0·01258	315
16	$O–NH_2$	−0·02381	595
16	$O–CH_4$	−0·03639	910
32	$S–O_2$	−0·01776	445
32	$S–CH_4O$	−0·05415	1352
32	$S–C_2H_8$	−0·09055	2262
28	$CO–N_2$	−0·01123	280
30	$NO–C_2H_6$	−0·04897	1225

It should be noted that some other doublets require higher resolutions to give adequate separation up to mass 600, notably H_2–D which can be expected to occur when deuterium labelling is used to clarify the structure of an organic sample. The maximum mass at which this doublet can be separated by an instrument with a resolution of 50,000 is less than 80 u.

1.2.4.2. Mass Measurement

In order to make a determination of the composition of one of the ion species at mass 28, the instrument must be able to resolve the different species and it must also allow the mass to be determined with an accuracy of at least 1 part in 2500. Mass measurements are normally made by comparing the mass of the unknown ion with that of a known 'reference' ion from a reference compound introduced with the sample. Such measurements could be made with a single focusing instrument by measuring the ratio of the accelerating voltages required to focus the two ion beams at the collector with a fixed magnetic field. The accuracy attained with a single focusing instrument can be high for molecular or parent ions but variations in the excess energy carried by fragment ions may give rise to mass measurement errors of the order of tens of p.p.m.

As in the case of resolution, the mass measurement accuracy required to identify the members of a doublet of given mass separation, increases with mass. In addition, the number of possible combinations of C, H, N and O increases rapidly with mass, from 4 at mass 28 to over 100 at mass 200. If sulphur is included this number is increased further and a mass measurement accuracy of even a few parts per million may be insufficient to determine the mass of an unknown ion uniquely so that further information from the breakdown pattern of the spectrum may be necessary to give a single solution.

These requirements in mass measurement accuracy and resolution can only be met by an energy focusing instrument.

The advantages of high resolution and accurate mass measurement in interpreting a spectrum are shown in *Figure 1.2.3.* In the low resolution spectrum of α-tetralone the masses of the peaks present are known only to the nearest whole number so that, unless the breakdown pattern can be recognized by comparison with a known sample, the composition of the parent and fragments may not be established without some additional information apart from the mass spectrum.

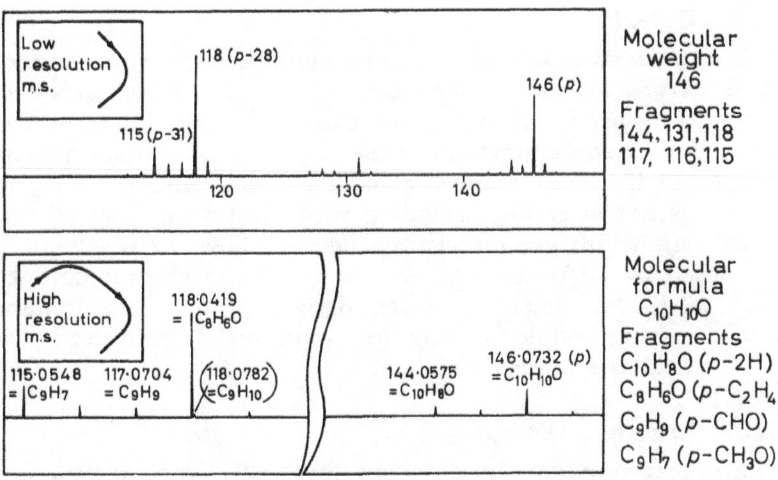

Figure 1.2.3. Comparison of low and high resolution spectra

In the high resolution case, the mass measurement of the parent clearly shows it to be $C_{10}H_{10}O$, and the fragment *p*–28 is largely *p*–C_2H_4 with a small peak due to *p*–CO.

Additional information on the routes involved in the breakdown pattern and so on the molecular structure of the parent, can be derived from metastable peaks in the low resolution spectrum, so it is desirable that the design chosen should give good sensitivity for metastable peaks and one should be able to change from low to high resolution operation quickly and easily.

1.2.4.3. Sensitivity

The basic sensitivity of the instrument is important also, since it is desirable to be able to obtain a recognizable spectrum at low resolution from a small quantity of sample material, of the order of a few nanogrammes. It is, therefore, desirable to cover the mass range up to at least 600 u at the full accelerating voltage.

The permissible slit widths, hence sensitivity, for a given resolution, increase as the linear dimensions of the instrument increase. In the case of the instrument to be described in detail, the AEI MS902, the radii of curvature in the 90° electric and magnetic field sectors are chosen to be 15 in. and 12 in. respectively, giving an overall size of the tube unit of 5 ft high, 4 ft wide and 6 ft long.

The maximum mass which can be tuned on the collector, when using the full accelerating voltage of 8 kV, is about 900 u.

1.2.4.4. Detector

The instrument is intended for organic analysis so it is assumed that either an electron bombardment or a field ionization source will be used, hence the total ion current output will not vary erratically over short periods of time, as is the case when a spark source is used. Simultaneous collection of all ion species in the spectrum is not essential, providing that the scanning speed and detector bandwidth are sufficiently high to allow the spectrum to be recorded in a few seconds, e.g. for gas–liquid chromatography (g.l.c.) work. A double focusing design, of the Nier–Johnson geometry, giving double focusing for one radius of curvature in the magnetic field was therefore chosen.

1.2.4.5. Practical Design

The design provides simultaneous first order angular (α), first order velocity (β) and second order angular (α^2), focusing. The last feature permits the use of a larger acceptance angle of ion beam and gives a small improvement in sensitivity The remaining second order aberration coefficients for $\alpha\beta$ and β^2 are small and, since the energy spread β from an electron bombardment source is also small, the sum of these aberrations is low enough to allow the theoretical resolution of the design to be approached quite closely.

Figure 1.2.4 shows a schematic diagram of the analyser system of the instrument.

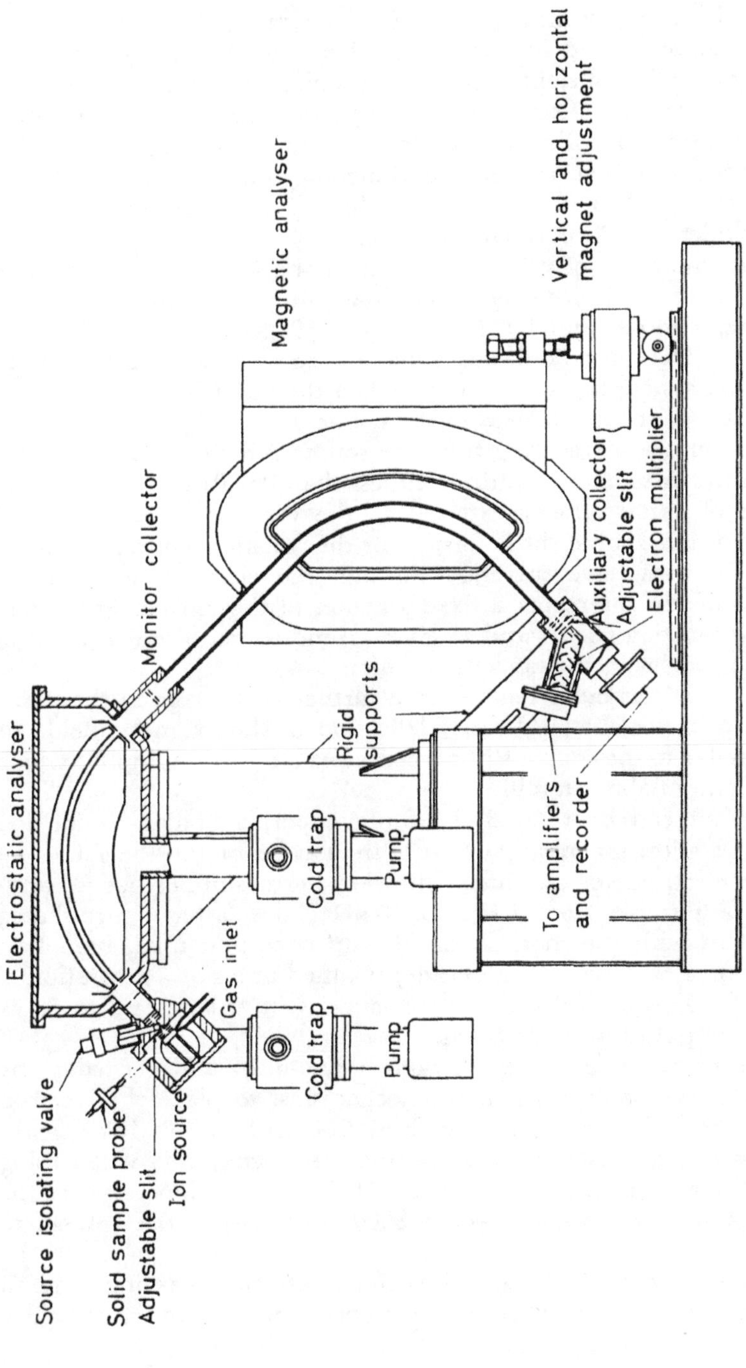

Figure 1.2.4. MS902 *schematic*

The ion source is arranged for good accessibility and easy removal of the filament and ion chamber; four inlet ports are provided, each giving line-of-sight access to the ionization region. One of the ports is normally fitted with a direct-introduction system which allows liquid or solid samples to be evaporated directly into the ionization chamber without breaking the source vacuum. Other ports are used for a conventional heated inlet system and g.l.c. direct inlet or helium separator systems.

Contamination of critical surfaces of the instrument by sample materials is minimized by differential pumping of the source and analyser regions, and by the provision of baking heaters for both regions. The source isolating valve is used to close off the analyser pumping system from the source, when the latter is removed either for cleaning or replacement of the filament.

The ion beam emerging from the source is defined by a slit (W_s) which is controlled in width, from less than 0·0001 in. to 0·01 in., by means of a micrometer external to the vacuum system. A real image of this source slit is formed in the 'monitor collector' region after the ions have travelled through the electrostatic analyser. This collector intercepts a fixed fraction of the ion current before mass separation occurs and so gives an indication of the amount of sample present in the ionization region.

Ions of a particular mass form a further real image in the plane of the collector slit (W_c) after deflection in the magnetic field. A slit, similar to the source slit, precedes the electron multiplier which has a gain of about a million.

The slit can be rotated, in a plane perpendicular to the ion beam, to allow for misalignment of the ion beam and the slit.

The electrometer amplifier following the multiplier has a bandwidth adjustable from 1 Hz to 10 kHz, the latter figure being consistent with the most stringent scan condition used, namely a mass range of 10 to 1 (one decade) scanned in 8 s at a resolution of 10,000[8]. The gain of the electrometer system is sufficient to allow single ion pulses to be detected.

The magnet can be moved both horizontally and vertically, by means of two calibrated screw mechanisms, to place the electron multiplier collector slit in the focal plane of the analyser fields. A mass range of 1–900 u can be covered magnetically when using the full ion accelerating voltage of 8 kV, and the upper end of the mass range can be extended to 3600 by reducing the voltage to 2 kV.

The magnetic field can be scanned at rates varying from a decade in mass in about 30 min to a decade in mass in less than 3 s.

A summary of the characteristics of the instrument is given in *Table 1.2.2.*

Table 1.2.2. Principal Features of MS902

R_e = 15 in., $\phi_e = 90°$, R_m = 12 in., $\phi_m = 90°$, $M_e = 0{\cdot}94$, $M_m = 0{\cdot}7$
M_m = linear magnification in the magnetic analysis
M_e = linear magnification in the electrostatic analysis

$$\text{Theoretical resolving power} = \frac{R_m\,(1 + M_m)}{2(W_c + M_e M_m W_s)}$$

$= 60{,}000$ with $W_s = W_c = 0{\cdot}1$ mil
(1 mil$=0{\cdot}001$ in. $= 25$ μm)

Ion accelerating voltage = 8 kV maximum
Electrostatic analyser voltage = ± 270 V maximum
Magnetic field = 12,000 gauss maximum
Mass measurement–by oscillographic peak matching or from high resolution scan
Ion detection (1) Electron multiplier (high resolution, fast scanning)
(2) Monitor collector (total ion current)
(3) Plate collector
Mass range–1–900 at 8 kV, 4–3600 at 2 kV

1.2.4.6. Mass Measurement

Precise mass measurement data can be obtained in two ways, either by simultaneously scanning the spectrum of the sample along with that of a known 'reference compound', at a high resolution and recording the peak centroids and times of occurrence[9], or by the so-called 'peak matching' method first used by Nier[10]. The choice of method is determined by the number of precise measurements required from the sample and the duration of the sample in time.

Scan Method—The conditions used in the scan method are a resolution of about 10,000 and a scan time of approximately 8 s per decade. The sample quantity consumed is less than 1 μg.

The analogue information, from the mass spectrometer, relating ion current intensity to time can be either recorded on magnetic tape for subsequent digitizing and processing by a computer in terms of peak intensity and mass, or processed directly if an on-line digitizer and computer is available. The masses of the peaks of the sample spectrum are determined with an accuracy of 2–5 p.p.m. by this method.

The data can be processed by the computer to give 'element maps' of the elemental formulae and intensity of the sample molecule and its fragments of the type originated by Biemann[11].

Figure 1.2.5 shows a simple element map from an androstene diketone derivative. The hetero-atom content increases from left to right and the mass increases downwards. The molecular formula is shown on the map in the position of greatest hetero-atom content and highest mass and is $C_{19}H_{26}O_2$. The number of dashes alongside the entry indicates the peak intensity on a logarithmic scale, hence the most intense peaks are $C_{19}H_{26}O_2$ and $C_8H_{12}O$.

Consideration of the formulae and intensity of the fragment peaks present in the element map can lead to the molecular

OF ANDROSTENE DIKETONE JOB NO 22868 5.4.66

SULPHUR CONTAINING FRAGMENTS IF ANY

	NO O0		NO O1		NO O2	
93	-----	C 7H 9				
95	-----	C 7H11				
97			------	C 6H 9		
104	----	C 8H 8				
105	------	C 8H 9				
106	----	C 8H10				
107	------	C 8H11	----	C 7H 7		
108	-----	C 8H12				
109	--	C 8H13	------	C 7H 9		
110			---	C 7H10		
111			----	C 7H11		
117	----	C 9H 9				
119	-----	C 9H11				
120	----	C 9H12				
121	-----	C 9H13	----	C 8H 9		
122	-----	C 9H14	----	C 8H10		
123			-----	C 8H11		
124			--------	C 8H12		
125			----	C 8H13		
128	----	C10H 8				
129	----	C10H 9				
131	---	C10H11				
133	----	C10H13	---	C 9H 9		
135	-----	C10H15	----	C 9H11		
136			------	C 9H12		
143	---	C11H11				
144	----	C11H12				
146	----	C11H14				
147	----	C11H15	----	C10H11		
148	------	C11H16	----	C10H12		
149	--	C11H17	-----	C10H13		
150			-----	C10H14		
159	----	C12H15	--	C11H11		
162			-----	C11H14		
163			----	C11H15		
171	---	C13H15				
173	----	C13H17	---	C12H13		
175			---	C12H15		
229			----	C16H21		
271					--	C18H23
286					--------	C19H26

Figure 1.2.5. Element map of androstene diketone derivative

structure, and knowledge of the elemental composition of all peaks allows one to follow the changes of hetero-atom content with the varying carbon numbers more easily than in the case where only the masses of a few peaks are accurately known. Nevertheless the element map shown is a simple one and complete interpretation of a more typical map can be time consuming, so it is reasonable to

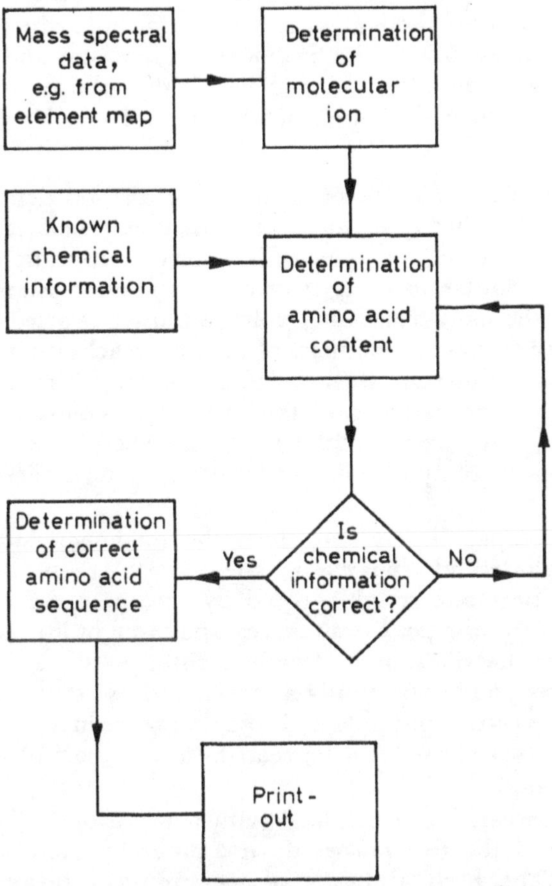

Figure 1.2.6. Logic diagram of computer programme for peptide analysis

expect that the computer can be used further to help in determining the structure from the fragmentation pattern.

The basic logic diagram of a computer programme for the analysis of linear peptides is shown in *Figure 1.2.6.*[12] In the first stage, the spectral data from the element map is fed in and the atomic composition of the molecular ion is calculated. Since the sample is a peptide it is known to be composed of amino acid groups and, in the second stage, possible combinations of amino acid groups are summed and are compared with the molecular ion formula in the loop shown. Any other known chemical information, for example from a qualitative amino acid analysis, can be introduced at this stage to reduce the number of combinations examined.

When the sum fits the known molecular formula, the amino acid content is compared with the fragment peak masses and intensities in order to determine the order in which the groups are lost from the molecule, hence the amino acid sequence can be determined and printed out.

Peak Matching Method—The peak matching method depends for its operation on the fact that, for a fixed magnetic field, the mass of the ions which fall on the collector is inversely proportional to their energy. In a double focusing instrument the electrostatic analyser voltage and the ion accelerating voltage must be varied in a given ratio in order to change the mass of the ions reaching the collector. The tolerance required of the accelerating voltage is reduced by the double focusing properties and the ratio of the masses of two ion species being compared is equal to the inverse of the ratio of the two electrostatic analyser voltages required to tune them onto the collector.

An oscilloscopic display, in which the unknown and reference peaks are plotted alternately on the screen, is used; the peak heights can be made roughly equal by means of the oscilloscope gain controls and the peak positions can be brought into coincidence by adjusting the dials of a decade resistor network. The final matching process may take only a few seconds, so that the effects of drift of the electric and magnetic fields are reduced. The ratio between the two masses can be read from the dials of the decade resistor network.

Mass ratios can be determined with a precision of 1 p.p.m. or 2 p.p.m. when the two masses do not differ by more than a few per cent. The high accuracy of measurement means that the number of possible elemental compositions which can be postulated for each peak measured is small.

Measurements can be made up to a 2:1 ratio, but with a precision of only about 10 p.p.m.

The maximum rate at which masses can be determined accurately, can approach one peak per minute provided that several unknown peaks are compared with the same reference peak.

In normal operation, when making use of the peak matching method, the spectrum of a sample is obtained first at a relatively low resolution, say one or two thousand, and recorded on the direct-writing galvanometer recorder. In general, mass measurements are required from the molecular ion, if this is recognizable on the low resolution spectrum, and three or four of the more intense fragment peaks. Hence the sample is reintroduced into the source after consideration of the low resolution spectrum along with a reference compound giving a small number of intense peaks at known masses in the neighbourhood of the unknown sample peaks and the mass comparisons can then be made with an accuracy up to 1 p.p.m. or 2 p.p.m.

This peak matching technique is clearly unacceptable for samples from which a more or less constant flow cannot be maintained for a period of 10–15 min, for example for normal on-line g.l.c. applications, in which case the scan method of mass measurement is used. Though element maps can be produced from data obtained by the peak matching method, this would clearly be too laborious for normal application.

1.2.4.7. Metastable Scanning

A further field in which the computer can simplify the interpretation of spectral data is in the use of metastable peak information in determining the fragmentation pathways in a molecule[13].

In a conventional low resolution scan, broad metastable peaks are seen at apparent mass $m_2{}^2/m_1$ from transitions:

$$m_1{}^+ \rightarrow m_2{}^+ + m_3$$

occurring in the region between the electrostatic and magnetic analysers.

Metastable transitions occurring between the source exit slit and the electrostatic analyser can be studied if the ratio of accelerating voltage to electrostatic analyser voltage is increased; in this case the normal ion spectrum disappears and only the results of metastable transitions are seen at the collector.

The transitions can be observed systematically if the fragment peak m_2 is tuned onto the collector with normal accelerating voltage. As the voltage is increased, the fragment peak m_2 disappears and other peaks due to transitions $m_1{}^+$ to $m_2{}^+$ reach the collector as the voltage passes through values $V_A \times m_1/m_2$.

If the accelerating voltage is measured at the centroid of the peak, then the mass m_1 of the precursor can be calculated. The main advantage of this method is that more metastable peaks can be detected, since the less intense metastables are not obscured by ions from the normal spectrum[14].

In general, one finds several precursors for each fragment, necessitating considerable calculation to determine the fragmentation paths. However, it is possible to provide digital data from the metastable scan to the computer and to programme the latter to calculate the paths and give a print-out of the form shown below. Twelve paths were found, in this case, from the fragmentation of mass 638.

```
DATE 17.11.66
TIME 06.35


COMPUTER ANALYSIS OF MASS SPECTRA


IDENTIFICATION OF FRAGMENTATION PATHWAYS    W.A.WOLSTENHOLME A/SA/09

638 → 494 → 437 → 338 → 310 → 267 →
638 → 437 → 338 → 310 → 267 →
638 → 338 → 310 → 267 →
638 → 269 →  86 →
638 → 409 → 337 →
638 → 328 → 271 →
638 → 302 → 245 → 213 →  72 →
638 → 203 →  72 →
638 → 620 → 310 → 267 →
638 → 623 → 494 → 437 → 338 → 310 → 267 →
638 → 494 → 338 → 310 → 267 →
638 → 494 → 269 →  86 →
```

The computer programme can be carried further by incorporating accurate mass measurement data and elemental composition of the fragment peaks. The computer print-out can now show the nominal masses of the fragment peaks in each path, together with the elemental composition of the unit lost at each stage.

1.2.5. INORGANIC ANALYSIS REQUIREMENTS

A first requirement for an instrument for inorganic analysis is that it should give approximately equal sensitivity for all elements; the only source which meets this requirement is the vacuum spark source[15], the characteristics of which, therefore, influence the instrument geometry chosen to a major extent.

Two rods of the sample material are mounted to form the electrodes of a vacuum spark in the ion source of the instrument. A high voltage of some 30–100 kV, normally of pulsed radio frequency, is applied across the electrode gap and ions representative of the sample material are formed directly in the spark discharge. It is a characteristic of the source that these ions have a wide energy spread, up to about 1 kV, so that a double focusing instrument is required to allow even a moderately high resolution to be obtained.

The actual position of the spark moves randomly about the tips of the electrodes so that the ion extraction efficiency of the ion accelerating plates varies in an erratic manner. As a result the total ion current is subject to quite short term variations over an order of magnitude. Consequently it is desirable to employ an instrument geometry giving double focusing for all masses simultaneously so that ion currents of different masses can be integrated simultaneously.

The spark discharge in the source gives rise to considerable local heating so that outgassing of the tips of the electrodes occurs, resulting in a rise in source pressure from the non-sparking level of 10^{-8} torr up to about 10^{-6} torr with some sample materials. The instrument must, therefore, have efficient differential pumping between the source and analyser regions to reduce the effects of gas scattering in the analyser.

This is particularly important when the instrument is used for the detection of very small concentrations of impurity in the major sample material; scattered ions of the major element could give rise to a background which could obscure the mass region of an impurity and worsen its detection limit.

1.2.5.1. Practical Design

The design chosen for the AEI MS702 is of a modified Mattauch–Herzog form and is shown in *Figure 1.2.7*.

The two electrodes of the sample material are preferably about ½ in. long and 0·01 in.^{2} in cross section and are mounted, in the source, on glass insulators with their tips approximately 0·02 in. apart. They can be moved, in three mutually perpendicular directions, by controls outside the vacuum system.

The positive ions formed in the discharge are accelerated, by the simple ion gun, through a potential difference of 20 kV. The final slit of this ion gun is 0·002 in. wide by 0·04 in. long and is the only interconnection between the source region and the first differential pumping chamber. This chamber precedes the electrostatic analyser and is pumped by a diffusion pump and cold trap system, independent of those of the source and the analyser regions. A differential pumping factor of 10,000, between source and analyser, is achieved in this way.

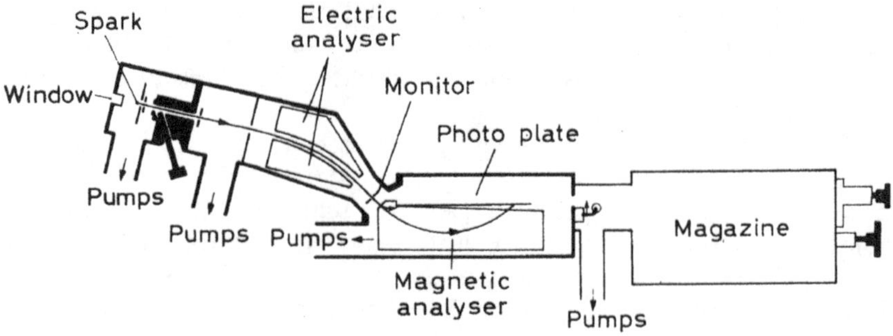

Figure 1.2.7. MS702 *schematic*

A monitor collector is mounted between the electrostatic and magnetic analysers and intercepts approximately half of the beam before it is mass-analysed. The monitor current serves as an indication of the ion current output of the spark and is used for setting up the spark conditions, while the integrated monitor current indicates the charge which has passed into the analyser region, being used to determine exposures. The largest mean ion current registered by the monitor collector is usually 10^{-9}A and the longest exposures, used in impurity detection work, represent a charge of 10^{-6} C and take $\frac{1}{2}$–1 h to record.

Ions leaving the monitor region enter the magnetic field, perpendicular to its boundary, and are deflected through almost 90° before leaving the exit boundary at a position dependent upon their mass. The particular modification of the Mattauch–Herzog geometry used, moves the focal plane beyond the magnetic sector field boundary and simplifies the engineering of the photoplate-moving mechanism. A 10 in. by 2 in. Ilford Q2 plate is normally used as the detector. The dimensions of the instrument are such that ions covering a mass ratio of 35:1 can be focused simultaneously on this length of plate. The magnetic field strength is usually set, for inorganic analysis, so that the mass range 7–250 is covered by the plate. The plate can be moved across the magnet gap by means of a lever mechanism operating through a bellows so that a series of graded exposures is usually recorded on one plate.

Up to 30 exposures can be accommodated, side-to-side, on one plate and up to eight plates can be stored, under vacuum, in the photoplate magazine at the rear of the analyser system. The magazine is pumped by a separate arrangement of diffusion pump and cold trap and can be isolated from the analyser system by both a sliding valve and a flap valve.

An exposed photoplate in the analyser can be returned to the magazine and replaced by an unexposed plate in less than 1 min and the process of venting the magazine, replacing the exposed plates with new ones and pumping the magazine to diffusion pump pressure, takes less than 15 min.

The techniques and application of the instrument to inorganic analysis, are described by Preston (see p. 239) so the techniques are not discussed further in this paper.

One of the requirements of impurity analysis already mentioned concerns the maintenance of clean vacuum conditions in the analyser region, which reduces gas scatter effects and allows detection limits of the order of one part in 10^9. An operating pressure of 10^{-8} torr is required in the analyser region for this purpose, and this is achieved by making high temperature baking possible. The vacuum system of the source and analyser regions is constructed of stainless steel and gold gaskets are used throughout; heaters are built onto the instrument so that it can be baked to a temperature of about 200°C in order to clean up the vacuum system periodically.

A system of hydraulic and electromagnetic vacuum valves enables the source to be completely isolated from the analyser, with the result that samples can be changed and a new analytical exposure started in a time as short as 10 min.

The principal features of the analyser system are shown in *Table 1.2.4.*

This design of instrument can also be used with other types of ion source, such as a Knudsen cell source and thermal ionization source for isotopic analysis of inorganic solids[16, 17], or an electron bombardment source for the analysis of organic materials.

The aberrations contributing to the ion beam image-width are smaller when an electron bombardment source is used, since the energy spread is small, hence the β^2 and $\alpha\beta$ terms are reduced. As a result higher resolutions can be achieved, near the optimum radius of curvature, than with a spark source. Resolutions of up to 30,000 have been obtained with the commercially available Mattauch design instruments with electron bombardment sources.

Either photographic or electrical detection methods can be used. In the electrical case, a twenty-stage electron multiplier is mounted behind an adjustable slit at the optimum focus near the high mass end of the focal plane. The spectrum is normally scanned by varying the magnetic field with fixed electrostatic analyser and ion accelerating voltages. Masses up to 700 u can be tuned onto the collector slit with an ion accelerating voltage of 10 kV.

Table 1.2.4. Principal Features of MS702

R_e = 15 in., ϕ_e = 31° 50′, R_m = 1·34–8·0 in., ϕ_m = 90°

Theoretical resolving power = $\frac{R_e}{W_s}$ approximately

= 7500 with W_s= 2 mil

(1 mil = 0·001 in. = 25 μm)

Ion accelerating voltage	=	20 kV maximum
Electrostatic analyser voltage	=	± 1 kV maximum
Magnetic field	=	17,000 gauss maximum
Ion detection	(1) Photographic plate (2) Monitor collector (3) Electron multiplier	
Ion source	(1) Spark (2) Electron bombardment	

Mass range 7–250 on one exposure at maximum magnetic field and maximum accelerating voltage

Pulse counting methods are also used on the output of the multiplier to give improved accuracy in isotope ratio analysis, or to extend the technique to the analysis of smaller sample quantities.

1.2.6. SUMMARY

Two designs of double focusing mass spectrometer have been considered from the viewpoint of two specific problems. These are, the analysis of organic vapours, solids and liquids, introduced through a conventional heated inlet system, or through a direct-introduction lock, or direct from a g.l.c. column, and the analysis of inorganic solids.

Other design solutions to these analytical problems are possible, but it is hoped that the practical solutions described serve as an example of some of the design philosophy used recently in mass spectrometry.

The author is indebted to his colleagues for their valuable advice, and to Associated Electrical Industries Limited for their permission to publish this paper.

REFERENCES

[1] BRUNNEE, C., *Mass Spectrometry*, REED, R. I. (Ed.), Academic Press, New York, 1964

[2] HERZOG, R. F. K. *Z. Phys.* 89 (1934) 447
[3] DEMPSTER, A. J. *Proc. Am. phil. Soc.* 75 (1935) 755
[4] BAINBRIDGE, K. T. and JORDAN, E. B. *Phys. Rev.* 50 (1936) 282
[5] NIER, A. O. and ROBERTS, N. R. *Phys. Rev.* 81 (1951) 508
[6] JOHNSON, E. G. and NIER, A. O. *Phys. Rev.* 91 (1953) 10
[7] MATTAUCH, J. and HERZOG, R. F. K. *Z. Phys.* 89 (1934) 786
[8] BANNER, A. E. *J. sci. Instrum.* 43 (1966) 138
[9] MCMURRAY, W. J., GREEN, B. N. and LIPSKY, S. R. *Analyt. Chem.* 38 (1966) 1194
[10] NIER, A. O., *Nuclear Masses and their Determination*, H. HINTENBERGER (Ed.), Pergamon Press, Oxford, 1957
[11] BIEMANN, K., BOMMER, P., DESIDERIO, D. *Tetrahedron Lett.* (1964) 1725
[12] BARBER, M., POWERS, P., WALLINGTON, M. J. and WOLSTENHOLME, W. A. *Nature, Lond.* 212 (1966) 5064
[13] BARBER, M. and ELLIOTT, R. M. *Am. Soc. Test. Mater. 12th A. Conf. Mass Spectrom. Allied Topics*, Montreal (1964)
[14] JENNINGS, K. R. *J. chem. Phys.* 43 (1965) 4176
[15] GORMAN, J. G., JONES, E. J. and HIPPLE, J. A. *Analyt. Chem.* 23 (1951) 438
[16] DICKENS, K. J. and POWERS, P. *Am. Soc. Test. Mater. 15th A. Conf. Mass Spectrom. Allied Topics*, Denver (1967)
[17] DICKENS, K. J. *Am. Soc. Test. Mater. 14th A. Conf. Mass Spectrom. Allied Topics*, Dallas (1966)

DISCUSSION

W. STECKELMACHER (EDWARDS HIGH VACUUM INTERNATIONAL)
What are the relative advantages of photographic versus electrical recording in high resolution instruments where either may be used as regards sensitivity?

G. A. ERROCK, *in reply.* Similar accuracies of mass and intensity measurement have been obtained with both methods using comparable quantities of organic sample.

1.3

A NEW CYCLOIDAL MASS SPECTROMETER

U. Scheidegger

1.3.1. HISTORY

THE first mass spectrometer, based on the cycloidal principle, was built by Bleakney and Hipple and reported in 1936 and 1938[1, 2]. These initial reports were followed by a series of publications on experiments and developments using different types and sizes of cycloidal analyser[3, 4]. An interesting although unsuccessful attempt has also been made to build a larger mass spectrometer of high resolving power based on the same principle[5]. For some time now a few commercially-built versions of the cycloidal mass spectrometer have been made available[6]. In the course of these developments further valuable information on the effect of non-uniform fields[7] and the occurrence of space charges[8] was gained.

1.3.2. THE CYCLOIDAL PRINCIPLE

The characteristic of such a design is the fact that the beam of ions is injected into crossed homogeneous magnetic and electrostatic fields.

If the total ion trajectory occurred in a magnetic field only, the ions would be deflected through 360° to form a circular path. Under the influence of the electric field only, they would be accelerated or decelerated in a linear trajectory. By superimposing the homogeneous electrostatic field at right angles to the homogeneous magnetic field, an environment is created in which the ions travel in a plane perpendicular to the two applied fields. They undergo a circular movement whilst being displaced linearly. The resulting trajectory is a trochoid (or cycloid). The shape of this path is dependent upon the relationship between the magnetic and electric fields; it can, in general, be a prolate or an oblate cycloid[9, 10]. Experiments by Bleakney and Hipple have shown, however, that an ion path describing a prolate cycloid is superior to the other in performance.

Under the conditions described above, with a given relationship between the two fields and a given focal distance (linear displacement of the source and the collector), ions of a predetermined mass leaving the source will follow a cycloidal trajectory and focus after a revolution of 360° (*Figure 1.3.1*). The ions are subject to direction as well as velocity focusing, giving double focusing, a characteristic of all cycloidal mass spectrometers.

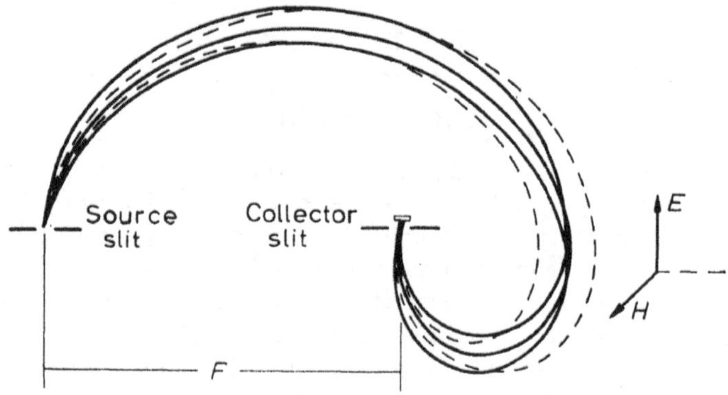

Figure 1.3.1. The ion trajectories in a cycloidal mass spectrometer of one particular m/e

The basic equation relating the various parameters for the cycloidal mass analyser is:

$$\frac{m}{e} = \frac{F}{2\pi} \frac{H^2}{E} \quad \ldots (1.3.1)$$

where H is the magnetic field strength, E the strength of the electric field and F the focal distance. Since neither the initial velocity nor the direction of motion of the ions appear in equation 1.3.1, it is evident that the initial values of these parameters will not affect the focus. Consequently, conditions for perfect focusing are also fulfilled by the cycloidal design.

Energy (velocity) focusing and angular (direction) focusing in the cycloidal mass analysis method are shown in *Figure 1.3.2.* In terms of the diagram, a lower energy represents a shorter secondary radius. In other words, a decrease in the accelerating voltage or a greater electric field will cause ions of the same mass to follow trajectory 1 rather than 2. For angular focusing, trajectory 3 in the diagram is formed by raising the plane on which the circle rolls so that it is parallel to, and above, the focal plane of the analyser.

Accordingly, trajectory 4 is obtained when the same plane is parallel to, but below, the focal plane of the analyser. This results in different angles or directions of motion with which the ions pass the source slit.

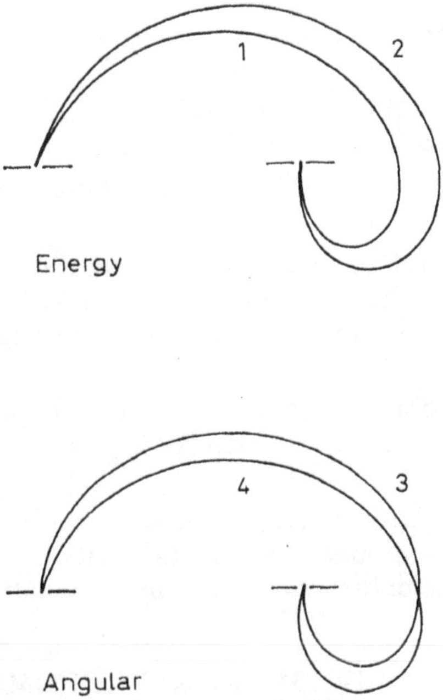

Figure 1.3.2. The influence of energy and angular spread on a cycloidal trajectory

In terms of different masses, increasing the magnetic field or decreasing the electric field will focus heavier ions. Scanning through the mass spectrum to bring the various ion beams of given m/e values through the collector slit and causing all ions to follow the same path, requires that both E and V (the ion energy or accelerating voltage) remain fixed while H is varied, or else that both H and the ratio of E to V remain constant while E is varied.

The advantages of a cycloidal mass spectrometer are manifold.

(*i*) The perfect focusing condition results in some very favourable consequences:

(*a*) Only two precise slits are required, one to define the position of the source and the other, placed at the focal point, to limit the mass range received.

(*b*) A relatively large fraction of the ions formed reach the collector. This results in a high sensitivity.

(*c*) In principle, a variety of sources can be used because any ions passing through the source slit will be focused at a distance depending only upon m/e and the field strengths.

(*ii*) The ions are contained completely within the homogeneous fields and do not pass through any fringe fields.

(*iii*) The shape and size of the cycloidal ion trajectory allows the design of a compact instrument.

The disadvantages of such an instrument are predominantly ones of mechanical considerations.

(*i*) The most serious limitation is the requirement that the magnetic field and the electrostatic field must be uniform over a larger volume than in any other type of mass spectrometer of the same performance.

(*ii*) The focal point of a given m/e is not unique in that ions of different m/e may pass through that focal point on the way to their own focus.

(*iii*) Negatively-charged ions which emerge from the source, will also be focused at the same point as the positive ions.

The last two difficulties can easily be eliminated in a practical design.

1.3.3. A NEW MASS SPECTROMETER OF CYCLOIDAL DESIGN

As illustrated in the previous subsection, the cycloidal principle offers many distinct advantages over other methods of mass analysis. It has therefore been used for the design of a new mass spectrometer, the M66 (*Figure 1.3.3*).

The aim was to build a routine instrument of medium resolving power mainly for the organic chemist, with special features such as ease of operation, precalibrated charts, linear read-out, etc. The results obtained demonstrate that the cycloidal design is, indeed, capable of meeting the necessary requirements. The design of the M66 and some applications are described in the following sub-sections.

1.3.3.1. Analyser

The performance requirements dictated the proper size of the analyser. When equation 1.3.1 is differentiated it can be seen that when E is constant, then $m/\Delta m = H/2\Delta H$ or that when H is constant, then $m/\Delta m = -E/\Delta E$. In order to achieve a resolution of, say, 4000 (50% peak width definition) the stability of H over the whole ion trajectory has to be 1 part in 8000 while the electric field has to be uniform to 1 part in 4000. To produce the necessary magnetic field a 9 in. low impedance magnet, which met all the criteria, was

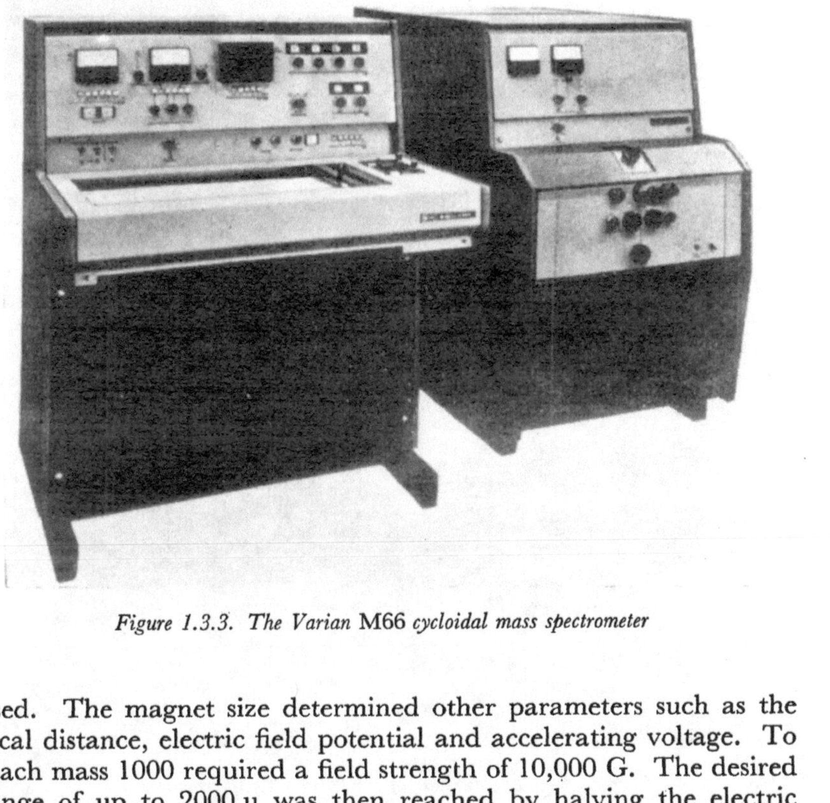

Figure 1.3.3. The Varian M66 *cycloidal mass spectrometer*

used. The magnet size determined other parameters such as the focal distance, electric field potential and accelerating voltage. To reach mass 1000 required a field strength of 10,000 G. The desired range of up to 2000 u was then reached by halving the electric field.

The analyser is an array of 16 rectangular frame electrodes, where each frame is operated at a slightly different electric potential. This establishes a region of uniform electric field E in the hollow interior of the frame electrode array. The latter is supported within a thin rectangular vacuum envelope (analyser chamber), which is located between the pole caps of the magnet. The chamber is operated under a vacuum of 10^{-6}–10^{-8} torr which is obtained by a VacIon™ pump. *Figure 1.3.4* shows the structure of the analyser. The ion source as well as the ion detector (Faraday cup) are mounted on the frame number 10, which has zero potential. The two are separated by the focal distance which amounts to 10 cm. The larger ring electrodes are negative with respect to ground potential (producing ion acceleration), and the six smaller rings are positive (causing ion deceleration).

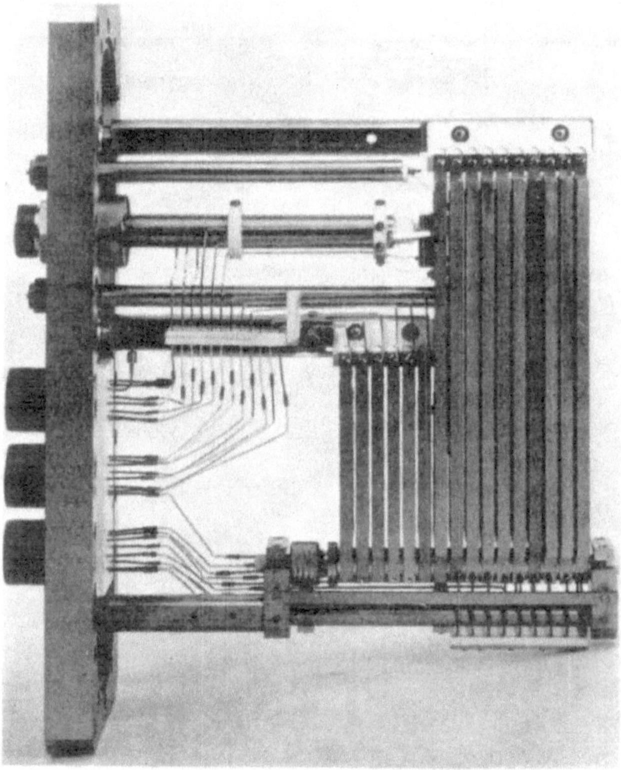

Figure 1.3.4. The analyser of the M66 *mass spectrometer*

With a given E and H only ions of a certain mass focus at the collector slit. The electrometer of the detector produces an output corresponding to the number of ions arriving. This output is amplified and fed to the y axis of an xy recorder. The signal is then recorded as a function of a scan of the magnetic field strength produced by the scan generator. In the cycloidal design, the ion trajectories are contained completely within the homogeneous magnetic field H and do not pass through any fringe fields. This allows the establishment of a very precise relationship between the mass of an ion and the magnitude of the magnetic field which brings it to the collector.

1.3.3.2. Ion Source

The electron impact source is located next to the analyser attached to the tenth frame electrode. Therefore, the source experiences the same magnetic field as the analyser, although the electric field in the source is different to the one in the analyser.

The source is designed to use two filaments to increase the time duration of maintenance-free operation. The rhenium filaments are located parallel to each other on opposite sides of the ionization chamber where they create an electron beam perpendicular to the ion path leaving the source. Each filament assembly can act either as a filament creating electrons, or as a collector trapping the electron beam. The electron energy, determined by the potential difference between filament and ionization chamber wall, can be regulated between 5 eV and 100 eV. The electron beam current may be set between 2 μA and 100 μA. The electrons, with a much smaller value of m/e than the ions, are collimated by the influence of the magnetic field. This influence, together with that of the electric field in the source, results in the electrons describing a small cycloidal path from the filament to the electron collector. The latter is positive relative to the ionization chamber walls to prevent a back flow of secondary electrons.

The vapours to be analysed enter the source in a direction which is parallel to the electrostatic field in the analyser. After electron impact the resultant ions are injected into the analyser by the electric field established between the back wall (repeller) and the source slit. This electric field within the source regulates the ion energy and may be varied between 50 V and 100 V. An ion collector, located just outside the source slit, traps a fraction of the ions entering the analyser and makes it possible to monitor the total ion current.

The source has been so designed that material can only be pumped from it through the source slit into the mass analyser and through the apertures for the ionizing electron beam. The resulting pumping speed in the source amounts to about 1 litre/s which is relatively slow and is an additional factor contributing to the high sensitivity of the M66.

Both the ion source and the analyser can be heated independently to 400° C and 250° C respectively. An automatic bake-out device is incorporated for the purpose of removing accumulated background material.

1.3.3.3. Sweep and Recording System

According to equation 1.3.1, in a cycloidal mass spectrometer m/e is proportional to the square of the magnetic field H. Thus a linear sweep of the magnetic field would yield a mass spectrum having a quadratic scale. A new sweep system has now been designed which eliminates this unfavourable mode of presentation and produces a linear mass read-out. It utilizes a modified Fieldial™ employing two Hall crystals. A block diagram of this sweep system is shown in *Figure 1.3.5.*

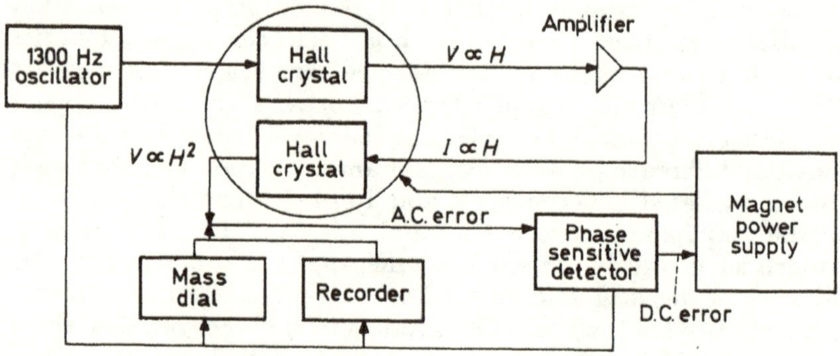

Cycloidal m.s. $M = \frac{kFH^2}{E}$

Hall effect $V_{out} = k' I_{in} H$

Figure 1.3.5. The M66 *sweep system to produce a linear mass read-out*

A constant a.c. power supply generates the current fed to the first Hall crystal. With the crystal properly oriented in the magnetic field, an output voltage is obtained which, according to the Hall effect, is proportional to the magnetic field. This voltage passes through an amplifier and is fed to the second Hall crystal whose output is now a voltage proportional to H^2. This signal, proportional to H^2, is then balanced in a summing transformer against a signal proportional to the mass generated by the sweep system (mass dial and recorder). The resultant a.c. error is phase-detected with a reference to establish a d.c. error. The latter is fed into the power supply which, in turn, regulates the magnetic field. The power supply responds to the d.c. error in order to maintain a field, so that the desired mass is brought to focus at the detector.

The mass generated by the sweep system, depends on the displacement of the recorder arm and the starting point of the sweep.

The starting point of the sweep is the mass set on the mass dial (a field-set mechanism which adjusts the magnetic field to focus a desired mass). The exact mass being traced at any given moment depends upon the sweep width chosen, the mass dial setting and the position of the recorder sweep arm.

This new sweep and recording system makes it possible to present mass spectra on precalibrated charts with the choice of a variety of sweep widths and any sweep offset (mass dial setting) between 10 u and 2000 u. Alternative ways of recording are possible, either on the oscilloscope or with an auxiliary recorder. With a direct-writing galvanometer recorder, read-outs up to a maximum rate of 100 u/s may be made, although with some reduction in linearity.

1.3.3.4. The Inlet and Vacuum System

The inlet system consists of a volatile sample inlet and a direct sample inlet for solids. The volatile sample inlet is a dual-inlet system. This allows one inlet to be cleaned while the other one is in use, the spectrum of a reference compound to be superimposed, or one system to be modified for special applications. Volatile samples are inserted by means of a small notch in the side of the cylindrical piston, which is then pushed through a sliding seal so that the notch, with the sample, moves into the sample chamber which is under vacuum. The sample then enters the source through the leak valve, which replaces the conventional fixed leak. With this new device it is possible to set the size of the opening so that the partial pressure of the sample in the source may be adjusted independently of the pressure in the sample chamber. The leak valve is set by observing a selected portion of the spectrum on the oscilloscope, the total ion current or the analyser pressure. The dual volatile inlet system may be heated to 200°C.

Solid samples are put into a melting-point capillary which is inserted in a well at the tip of the direct sample introduction probe and held in place by a small nozzle. The probe is then inserted through the sliding seals of the inlet. When the probe is next to the ion source the sample is approximately 5 mm from the ionizing electron beam. The sample in the direct introduction probe is then heated until the temperature is reached (with an upper limit at 500°C), which gives the desired pressure of the material in the source.

The pumping system of the M66 consists of three pumps which make it possible to pass from atmospheric pressure to the 10^{-8} torr range. The analyser chamber, which houses both the analyser and the ion source, is pumped by a VacIon™ pump, which is a sputter ion pump, and works from 10^{-4} torr to 10^{-8} torr or lower. The two remaining pumps, namely the VacSorb™ in series with a two-stage, mechanical pump, are used to prepump the whole system to pressures where the VacIon™ begins to operate during the initial pump-down operation and to evacuate the entire inlet system during normal operation. The VacSorb™ acts as a foreline trap in conjunction with the mechanical pump when no liquid nitrogen is used. The system will then reach the limit of the mechanical pump in the 10^{-4} torr range. It may also be operated with liquid nitrogen as a cryogenic pump. In this case a faster pumping speed is achieved and a better intermediate vacuum obtained (within the 10^{-5} torr range). The fact that the VacSorb™ pump is positioned between the mechanical pump and the inlet and high vacuum system prevents back-diffusion of any pump oils into the mass spectrometer system.

1.3.3.5. Resolution, Sensitivity and Precise Mass Determination

In mass spectrometry both the resolution (ρ) and sensitivity (S) factors are interrelated. To a first approximation the relationship between the two is $S \sim 1/\rho$. Sensitivity is related to the number of ions admitted from the source to the mass analyser, whereas resolution depends on the width of the ion beam admitted into the mass analyser. In other words, a narrow beam is desired to maximize resolution while a large number of ions are required to achieve maximum sensitivity.

The performance limits on the resolution in cycloidal mass spectrometers can be expressed by the following equations:

$$\rho = F/S_{\text{max.}} \quad \text{(geometrical limit)}$$
$$\rho = 2\Delta H/H$$
$$\rho = \Delta E/E$$

where ρ is the resolution ($m/\Delta m_{\frac{1}{2}\text{max.}}$) and $S_{\text{max.}}$ is the width of the widest slit.

It should be noted that with the above limits the influence of abberations, i.e. field inhomogeneities, etc., are not considered. The perfect double focusing scheme of the cycloidal design provides a constant resolution throughout the whole mass range, which is not true for sector type instruments. With the aid of variable source and collector slits, the resolution $m/\Delta m_{\frac{1}{2}\text{max.}}$ of the M66 can be varied between 1000 and 4000. With fixed slits a resolution of better than 2000 is obtained. (The approximate value for the 10% valley definition is obtained by dividing the $m/\Delta m_{\frac{1}{2}\text{max.}}$ data by a factor of 2.)

The sensitivity of a mass spectrometer system is a combination of many factors. The three most directly related are the design of the source upon which the rate of usage of material is dependent, the efficiency with which ions are transmitted from the source to the detector for a given resolution and the detector sensitivity. In mass spectrometry the sensitivity is stated in many ways. One way is to define it as the ion current per unit pressure in the source. For the M66 this value is greater than or equal to 10^{-5} A/torr {for $\rho \geq 2000$ and the ion electron beam (70 V) set at greater than 75 μA}. Another way is to express it in terms of sample usage. This value was calculated to be $3 \times 10^{-1}\sqrt{\text{mW}}$ ng/s with a standard source pressure of 10^{-6} torr. This rate of usage will give a signal-to-noise ratio of greater than 500 to 1 on a base peak representing 20% of the total ion current. *Figure 1.3.6* shows the mass spectrum obtained from 1 μg of a steroid of molecular weight 274.

Precise mass measurement requires a mass analysis method which handles both angular and energy effects and the ability to locate peak centres on a known and linear mass scale. These requirements are met by the perfect double focusing capability of the cycloidal scheme and the additional linear scale incorporated in the M66. This linear scale is precise when a reference peak of known mass is contained on the same scale. With the M66 a precise mass measurement accuracy of 10 p.p.m. is obtained if the standard and unknown peaks are separated by not more than 2·5 u. This accuracy can be increased still further by the use of repetitive measurements.

1.3.3.6. Metastables

Robinson[11] and Beynon[12] are the only authors who have, in the past, reported preliminary studies on metastable ions in cycloidal mass spectrometers. Recently, Benz and Brown[13] have performed further investigations in this field and have presented a mathematical model that explains in detail the line shapes resulting from metastable ion transitions in a cycloidal instrument.

There is a principal difference between the decay of metastable ions in a sector and in a cycloidal mass spectrometer. In a sector instrument, the positions, on the mass scale, of ions resulting from metastable ion decays, are related to a momentum analysis for decay occurring in the source or in the field-free region between the source and the magnetic sector. In the cycloidal mass analyser, however, the distribution of the daughter ion intensities on the mass scale is related to the time of decay when the ions are on their

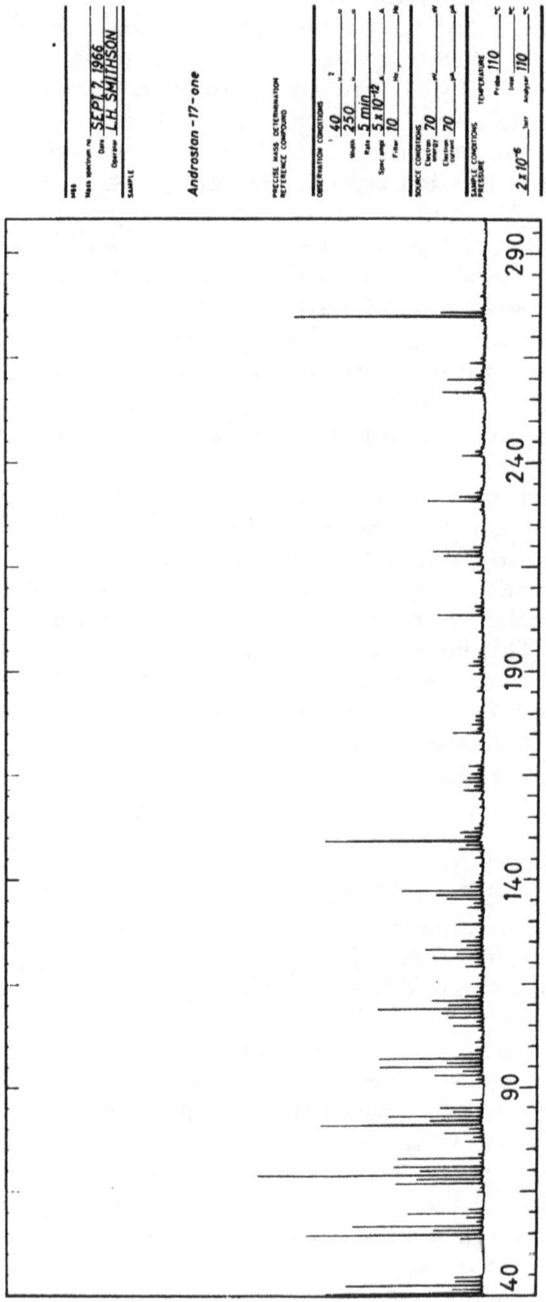

Figure 1.3.6. Mass spectrum of 1 μg of androstan-17-one

cycloidal trajectory. Using the model for cycloidal instruments given by Benz and Brown, it is possible to identify both the metastable parent ion and the daughter ion resulting from this metastable decay, as well as to determine the half lives of metastable ions in the range 10^{-7} to 10^{-5} s.

1.3.3.7. Combination with Gas Chromatography

The combination of gas chromatographic techniques with mass spectrometry has become exceedingly useful, because the technique, which permits scanning the entire mass spectrum of a component while it emerges from the gas chromatograph, can be an ideal solution to many structural problems. Although the two analytical methods have many facets in common, there is one difficulty to overcome, namely the extreme difference in pressure between the two instruments. This difference can best be reduced with the aid of an interface between the two instruments, which attenuates the carrier gas to an acceptable level, while at the same time transferring the organic material to the mass spectrometer with as small a loss as possible. Several such devices have been built and are commercially available[14, 15]; all make some use of the more rapid gaseous diffusion rates of light gases with respect to heavier ones.

Recently a new interface has been developed by Llewellyn[16]. The concept used is a new one. In essence the interface consists of a thin membrane, normally an elastomer, which is permeable to organic vapours. The basic property of this membrane is that it acts rather like a liquid phase. The carrier gas and the organic vapours come into contact with it and partition occurs. The permeability of the membrane is small for the carrier gas, because of its low solubility, despite the fact that its diffusion is rapid. For the organic vapour, the large solubility results in a large permeability and a higher percentage of organic vapour passes to the low pressure side of the membrane. In other words, this separator takes advantage of the different solubilities between carrier gas and organic vapours in the membrane, and uses the same principle as the gas chromatographic separation. The only difference is that material passes through a layer of 'liquid phase' instead of going back into the carrier gas stream.

In order to achieve better vacuum in the mass spectrometer and to further enrich the organic material, two membranes are connected in series to give a two-stage molecular separator. With this system an extremely low helium flow into the mass spectrometer of less than 10^{-6} torr l/s is obtained. The enrichment factor for the organic material is very large and is at least 10^{5} when helium, nitrogen, hydrogen or argon are used as carrier gases, with typical flow rates of up to 50 ml/min. The sample recovery is approximately 50%.

REFERENCES

[1] Hipple, J. A. and Bleakney, W. *Phys. Rev.* 49 (1936) 884

[2] Bleakney, W. and Hipple, J. A. *Phys. Rev.* 53 (1938) 521

[3] Monk, G. W., Graves, J. D. and Horton, J. L. *Rev. scient. Instrum.* 18 (1947) 796

[4] Monk, G. W. and Werner, G. K. *Rev. scient. Instrum.* 20 (1949) 93

[5] Mariner, T. and Bleakney, W. *Rev. scient. Instrum.* 20 (1949) 297

[6] Robinson, C. F. and Hall, L. G. *Rev. scient. Instrum.* 27 (1956) 504

[7] Robinson, C. F. *Rev. scient. Instrum.* 27 (1956) 509

[8] Robinson, C. F. *Rev. scient. Instrum.* 27 (1956) 512

[9] Kiser, R. W., *Introduction to Mass Spectrometry and its Applications*, Prentice-Hall, New Jersey, 1965

[10] Beynon, J. H., *Mass Spectrometry and its Application to Organic Chemistry*, Elsevier, Amsterdam, 1960

[11] Robinson, C. F. *Rev. scient. Instrum.* 27 (1956) 507

[12] Beynon, J. H., *Mass Spectrometry and its Application to Organic Chemistry*, Elsevier, Amsterdam, 1960

[13] Benz, H. P. and Brown, H. W. *J. chem. Phys.* in press

[14] Ryhage, R. *Analyt. Chem.* 36 (1964) 759

[15] Watson, J. T. and Biemann, K. *Analyt. Chem.* 36 (1964) 1135

[16] Llewellyn, P., Papers presented at *Pittsburgh Analyt. Conf.* (1966), *Int. Conf. Gas Chromat.*, Rome, (1966), *Unpublished work* Varian Technical Information Quarterly (1966)

DISCUSSION

W. Snedden (St. Bartholomew's Hospital)
What is the approximate pressure in the ionization chamber? Is there any evidence of ion–molecule reactions at the pressure used?

U. Scheidegger, *in reply.* The pressure in the ionization chamber is about 10 times the pressure in the analyser, this means normally between 10^{-5}–10^{-6} torr. We have not observed any evidence of ion–molecule reactions.

W. A. P. Young (Atomic Weapons Research Establishment)
How stable is the field measurement against temperature? Do you have to select Hall crystals carefully?

U. Scheidegger, *in reply.* Field measurement is found to be stable with temperature, over the temperature range of the analyser of up to 250° C. Hall crystals are carefully selected and tested with the Fieldial™. Hall crystal probes, however, are exchangeable.

G. F. Snow (Science Research Council)
What happens to the organic material in the sputter ion pump?

U. Scheidegger, *in reply.* The gas molecules are trapped permanently in the pump, by the formation of chemically stable compounds and by ion burial. When continuously operated, this pump may be expected to function without any maintenance for periods in excess of 1 year. The electrodes can easily be replaced.

G. W. BALL (20th CENTURY ELECTRONICS)
Do you use sputter ion pumps when monitoring g.l.c. eluents? How do you cope with the viable gases?
U. SCHEIDEGGER, *in reply*. Yes, the VacIon™ pump can easily handle the quantity of carrier gases introduced into the spectrometer when the Llewellyn separator is used as the interface. This is also true for argon although the pumping speed for argon is 10 times slower than for nitrogen, or helium.

W. SNEDDEN (ST. BARTHOLOMEW'S HOSPITAL)
In the early cycloidal focusing instruments, a fairly elaborate procedure was required to adjust the electrostatic field plates for optimum performance. Has this been overcome in the Varian design?
U. SCHEIDEGGER, *in reply*. The potential drop between each field plate is exactly the same. Provision is made for adjustment of the first and last plates to overcome end effects.

1.4

A TIME OF FLIGHT MASS SPECTROMETER

P. R. Wadham

1.4.1. INTRODUCTION

A TIME of flight mass spectrometer is a device in which ions of differing masses are given the same energy and are allowed to traverse a field-free space. Because of their differing velocities, the lightest ions arrive first at the end of the drift space where there is a collector. The amplified current signal is displayed on an oscillo-scope. In the spectrometer under discussion, the ions are formed by a thin electron beam and are pulsed into the drift space which is earthed (*Figure 1.4.1*).

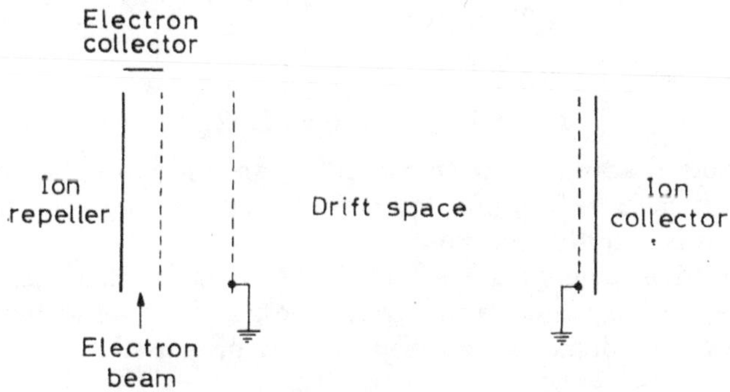

Figure 1.4.1. Simple arrangement of electrodes

In 1965, Elliott Brothers (London) Ltd. started to develop a time of flight mass spectrometer which would resolve masses up to 40 u, show detectable peaks up to 100 u and have a sensitivity of 10^{-11} torr. Work started on the development in April 1966 and by the end of 1966 a working model was demonstrated.

1.4.2. HISTORY

The first time of flight mass spectrometer was proposed by Stephens[1] in 1946. Shortly afterwards, Cameron and Eggars[2] were able to demonstrate crude mass spectra from what they called an 'ion velocitron'. Keller[3], and Wolff and Stephens[4] described similar machines in 1949 and 1953, but the most significant advance came, in 1955, with the time of flight mass spectrometer described by Wiley and McLaren[5]. With a two-field ion extraction system and a very wide-band electron multiplier, they achieved adjacent mass separation up to 200 u.

1.4.3. ADVANTAGES AND DISADVANTAGES

The main advantage of a time of flight mass spectrometer over other types, is the speed at which the complete spectrum can be displayed.

Applying the equation $eV = \frac{1}{2} mv^2$, gives the velocity v of a particle to be $\sqrt{\frac{2eV}{m}}$. For a proton, $e/m = 9\cdot57 \times 10^7$ C/kg therefore:

$$v = 1\cdot38 \times 10^4 \times v^{\frac{1}{2}} \text{ m/s} \qquad \ldots (1.4.1)$$

For an accelerating voltage of 1000 V and a drift tube length of 0·5 m, the transit time t, of a proton, will be:

$$t = \frac{0\cdot5}{1\cdot38 \times 10^4 \times 10\sqrt{10}} = 1\cdot15\ \mu\text{s} \qquad \ldots (1.4.2)$$

and, for a mass of 200 u:

$$t = 1\cdot15 \times \sqrt{200} = 16\cdot2\ \mu\text{s}$$

A complete scan, up to mass 200, can therefore be made in 16·2 μs, which is very useful for monitoring a system in which the gas content is rapidly changing.

The disadvantage of a time of flight mass spectrometer is the limited resolution which is inherent unless a very broad band, and hence expensive detector and display system, is used.

1.4.4. DESCRIPTION

1.4.4.1. The spectrometer

The spectrometer is built into a stainless steel cross-piece (*Figure 1.4.2*). On one arm of the cross is an ionization gauge and bakable leak and on the opposite arm is the field-free drift tube and an electron multiplier. The electrode system is suspended from the top plate, which has lead-through connections for the various electrodes.

The directly-heated cathode is made from 0·010 in. diameter tungsten wire, wound on a 0·080 in. diameter molybdenum mandrel, which has been ground to a D section in order to give a plane emitting surface. The filament runs at 40 W and a pulse current of 200 mA is achieved with a duty cycle of 0·003. The electron beam is 2 mm thick by 15 mm wide and is modulated by pulsing a focus electrode which is normally held at cathode potential. When the

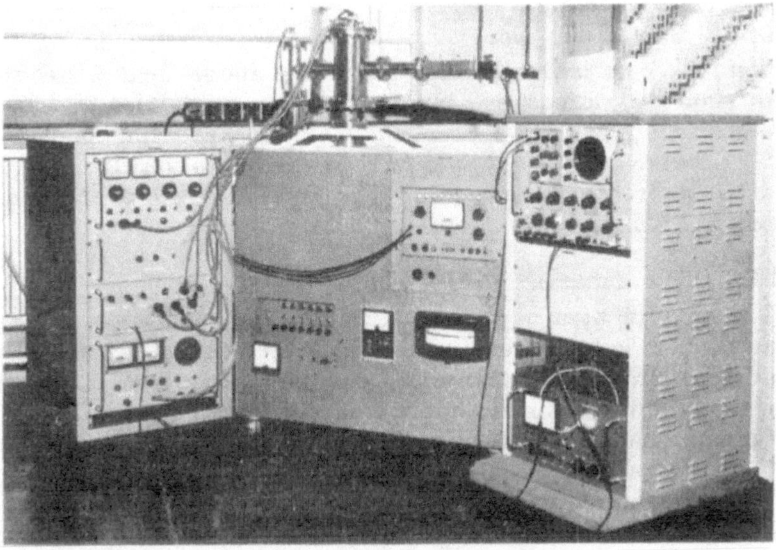

Figure 1.4.2. Photograph of apparatus showing, from left to right, power supplies, spectrometer and pump unit, and display console

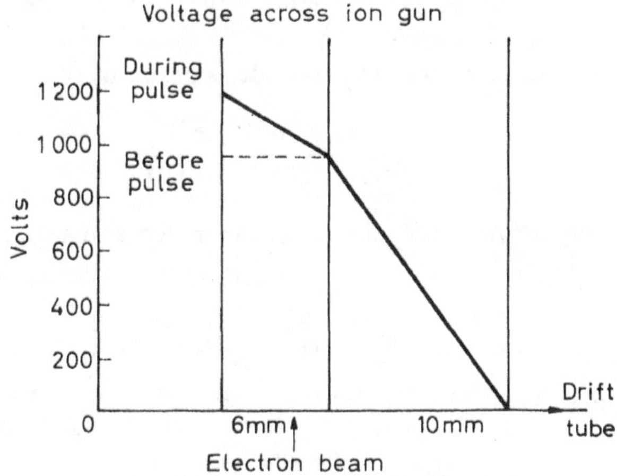

Figure 1.4.3. Ion accelerating field

electron pulse is switched off, a pulse is applied to a repeller plate, situated parallel to the electron beam. Ions, which were formed by the electron beam, are thus accelerated, away from the repeller plate, through an accelerating field (*Figure 1.4.3*) and an earthed grid, into the field-free drift space. They are eventually accelerated into the dynode of an electron multiplier.

1.4.4.2. Power supplies

Besides the variable d.c. supplies for the anode, heater, electron multiplier and various grid biases, there is a modulator to deliver two positive pulses of varying length up to 6 μs with up to 6 μs delay between them. The first pulse is applied to the electron gun and the second to the ion gun. The repetition rate of this pulse system is 1 kHz.

1.4.4.3. The detector

The detector used at present is a Venetian blind-type electron multiplier, in conjunction with a high speed oscilloscope. The multiplier limits the achievable resolution, since the transverse length of the first dynode is approximately 3 mm.

Suppose that the limit of resolution occurs when the distance between two adjacent peaks is equal to the width of one of them, and that this occurs at mass m.

Then, considering an infinitely thin sheet of ions of mass m, from equation 1.4.1, their speed is given by:

$$v = \frac{1{\cdot}38 \times 10^4 \times 10\sqrt{10}}{\sqrt{m}} \; m/s$$

The time taken to cross the first dynode is given by:

$$\Delta t = \frac{3 \times 10^{-3}\sqrt{m}}{1{\cdot}38 \times 10^4 \times 10\sqrt{10}} \; s$$

Then, from equation 1.4.2, the time taken for a mass m to reach the first dynode is:

$$t = \frac{0{\cdot}5\sqrt{m}}{1{\cdot}38 \times 10^4 \times 10\sqrt{10}} \; s$$

Similarly, for a mass $(m-1)$, the time taken is:

$$t' = \frac{0{\cdot}5\sqrt{m-1}}{1{\cdot}38 \times 10^4 - 10\sqrt{10}} \; s$$

Therefore, for the limit of resolution:

$$t - t' = \Delta t$$

$$3 \times 10^{-3} \sqrt{m} = 0{\cdot}5 \sqrt{m} - 0{\cdot}5 \sqrt{m - 1}$$

from which $m = 83.$

To improve the resolution, t could be made smaller by a different choice of multiplier or $(t - t')$ could be made larger by lengthening the drift tube. The latter tends, however, to lower the sensitivity.

1.4.5. FOCUSING

Because the electron beam is not infinitely thin and the ions have a finite energy before the pulse is applied, variable pulse height and delay are introduced to achieve spatial focusing and initial energy focusing. In practice, it is found that the complete spectrum can be displayed, but that final adjustment of these two parameters is necessary when observing any group of ions at their optimum resolution.

1.4.6. RESULTS

Figure 1.4.4 shows the background scan of the unbaked system up to 45 u. The pressure in the system was 10^{-6} torr. The horizontal scale is 0·1 μs/cm.

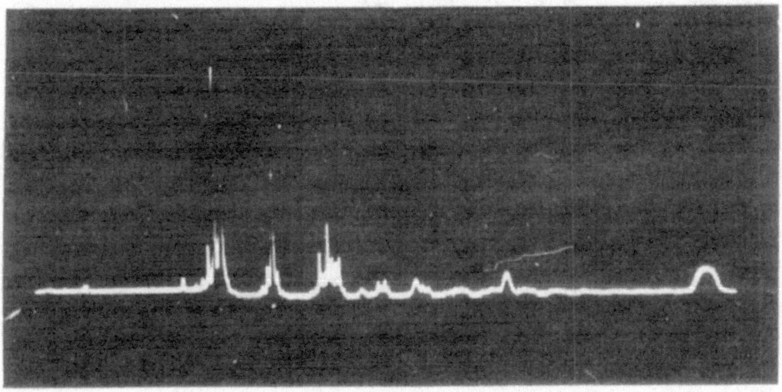

Figure 1.4.4. Photograph of background scan showing H_{21}^{+} *on the left, through to* H_{8}^{+} *on the right*

Figure 1.4.5 shows the mass scan of mercury. Although the peaks are not resolved, the isotopes, and also their relative abundance, can be clearly seen.

The machine has proved very reliable and simple to use. Filament lives of several hundred hours have been noted and, in event of filament failure, the fitting of a new filament assembly is easily achieved with normal laboratory tools.

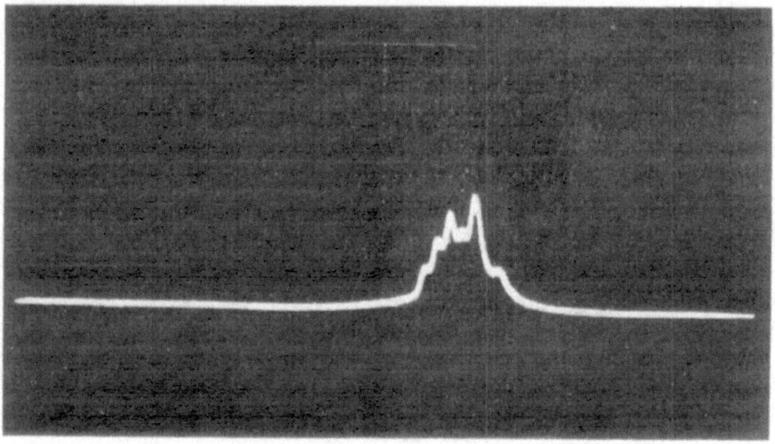

Figure 1.4.5. Mass scan of mercury

REFERENCES

[1] STEPHENS, W. E. *Phys. Rev.* 69 (1946) 691
[2] CAMERON, A. E. and EGGARS, D. F. *Rev. scient. Instrum.* 19 (1948) 605
[3] KELLER, R. *Helv. phys. Acta* 22 (1949) 386
[4] WOLFF, M. M. and STEPHENS, W. E. *Rev. scient. Instrum.* 24 (1953) 616
[5] WILEY, W. C. and MCLAREN, J. H. *Rev. scient. Instrum.* 26 (1955) 1150

DISCUSSION

W. SNEDDEN (ST. BARTHOLOMEW'S HOSPITAL)
Has the Bendix magnetic electron multiplier been tried on this instrument?

P. R. WADHAM, *in reply*. No, it has not. It would improve the performance but at considerable expense.

W. SNEDDEN (ST. BARTHOLOMEW'S HOSPITAL)
How reproducible are the spectra obtained at the maximum scan rate?

P. R. WADHAM, *in reply*. At 1 kHz, the fixed scanning rate of the instrument, spectra are reproducible over several hours.

J. W. COLEMAN (BRITISH PETROLEUM CO.)
What is the detection limit of the instrument?

P. R. WADHAM, *in reply*. Using the existing set-up, this is quoted as 10^{-10} torr/mV. With an improved oscilloscope, this could be reduced 50 times and with the Bendix multiplier this can possibly be reduced by a factor of 100.

J. F. J. TODD (UNIVERSITY OF KENT)
Is it bakable?

P. R. WADHAM, *in reply*. Yes. The baking temperature is dictated by the type of multiplier used.

W. STECKELMACHER (EDWARDS HIGH VACUUM INTERNATIONAL)
What is your experience of multiplier gain stability with time and following baking.

P. R. WADHAM, *in reply*. Stable for about 1 week and then reducing slowly over a period of 6 months, the gain reducing from 10^5 to 5×10^3. Working pressure and substances used will influence this.

1.5

SOME RECENT AND PROPOSED DEVELOPMENTS IN THE FIELD OF QUADRUPOLE MASS SPECTROMETERS AND THEIR APPLICATION AS VERSATILE MASS ANALYSERS

G. W. Ball

1.5.1. INTRODUCTION

THE head of the quadrupole mass spectrometer is shown schematically in *Figure 1.5.1.* The primary components are the ionizer, the quadrupole section and the ion detector. The substance, which is to be analysed, is introduced into the ionizer as a gas at low pressure. A small percentage of the atoms or molecules which make up this substance, is ionized by electron bombardment. These ions are then accelerated and focused into the quadrupole section. The ion

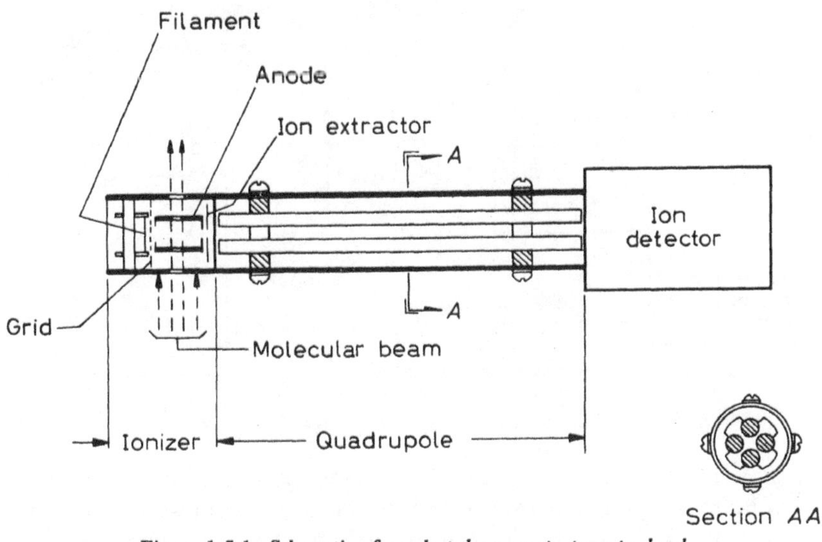

Figure 1.5.1. Schematic of quadrupole mass spectrometer head

beam is filtered by permitting only those ions within a specific range of e/m ratios to pass through the quadrupole. (The mechanism of this filtering action is discussed in a later section.) Those ions which are able to pass through the filter are collected by the ion detector which is usually an electron multiplier or a Faraday cup.

The output current of the detector is a measure of the number of atoms or molecules in the substance which, when ionized, have a particular e/m. The e/m ratio detected is determined from the values of a d.c. voltage and an r.f. voltage which are applied to the poles of the quadrupole section. The majority of the ions produced in the ionizer, are singly charged, therefore the mass of the atoms or molecules may be directly determined. A display of quantitative abundance as a function of atomic mass weight can be conveniently presented on many types of instrument, for example, oscilloscopes, *xy* recorders and strip-chart recorders.

1.5.2. THEORY OF OPERATION

The heart of the r.f. quadrupole mass spectrometer is the quadrupole section. This assembly is composed of four stainless steel rods, approximately 5 in. long and $\frac{1}{4}$ in. in diameter. These rods are

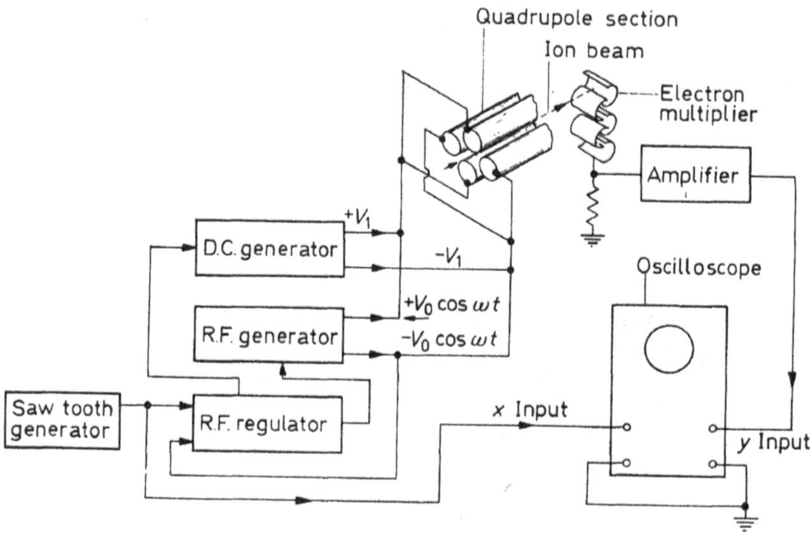

Figure 1.5.2. Block schematic of d.c. and r.f. supplies to rods

precisely located, in a rectangular array, by alumina insulators. A d.c. voltage and superimposed r.f. voltage, are applied to the rods as shown in *Figure 1.5.2*. An electrostatic field is thereby generated in the region between the rods. If the rods were hyperbolic cylinders (*Figure 1.5.3*) rather than circular cylinders, the potential of this electrostatic field would be:

$$\phi = (V_1 + V_0 \cos \omega t)\,(x^2 - y^2)/r_0^2 \qquad \ldots (1.5.1)$$

This potential is approximately that which is generated in the actual geometry.

When a singly-charged ion enters the filter, it is accelerated by the electrostatic field. The cartesian force components are:

$$F_x = -e\frac{\partial \phi}{\partial x} = -e(V_1 + V_0 \cos \omega t)\,(2x/r_0^2) \qquad \ldots (1.5.2)$$

$$F_y = -e\frac{\partial \phi}{\partial y} = +e(V_1 + V_0 \cos \omega t)\,(2y/r_0^2) \qquad \ldots (1.5.3)$$

$$F_z = -e\frac{\partial \phi}{\partial z} = 0 \qquad \ldots (1.5.4)$$

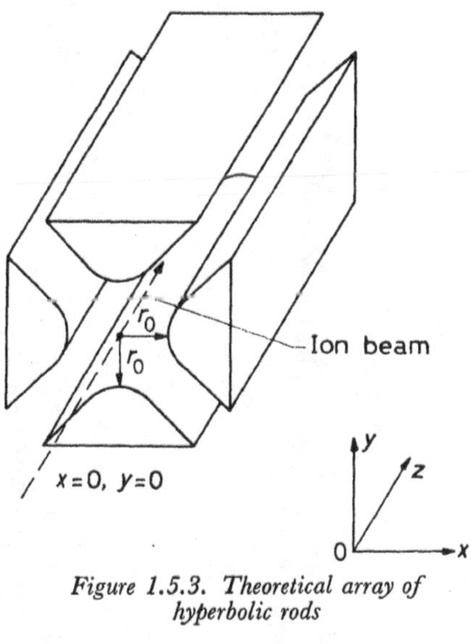

Figure 1.5.3. Theoretical array of hyperbolic rods

where e is the electronic charge. Therefore the equations of motion of a singly charged ion mass, m, are:

$$m\ddot{x} + (2e/r_0^2)\,(V_1 + V_0 \cos \omega t)\,x = 0 \qquad \ldots (1.5.5)$$

$$m\ddot{y} - (2e/r_0^2)\,(V_1 + V_0 \cos \omega t)\,y = 0 \qquad \ldots (1.5.6)$$

$$m\ddot{z} = 0 \qquad \ldots (1.5.7)$$

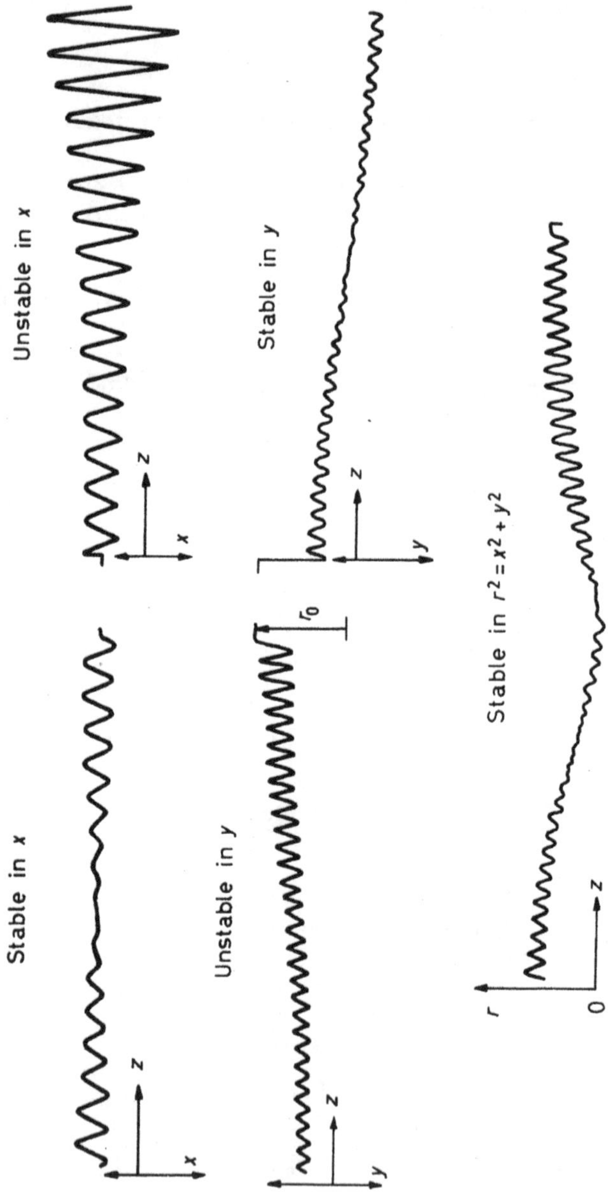

Figure 1.5.4. Typical ion trajectories

Equation 1.5.7 may be integrated to determine the axial motion of the ion, i.e.:

$$\dot{z} = \dot{z}_0 = \text{constant} \qquad \ldots (1.5.8)$$

Therefore the axial velocity of any ion is constant and is not affected by the voltages applied to the poles of the filter. The magnitude of the axial velocity is its value at the entrance to the filter. The ionizer system thereby governs the time required for an ion to pass through the filter.

The filtering action of the quadrupole section results from the trajectory characteristics of equations 1.5.5 and 1.5.6. Under certain conditions the amplitude of the trajectory increases without bound; the ion then collides with one of the poles and is removed from the ion beam. The characteristics of these Mathieu equations have been analysed by many investigators[1] and, for the particular case of the r.f. quadrupole mass spectrometer, stability diagrams have been developed by Paul[2], Stanford Research Institute[3] and Electronic Associates, Inc. (E.A.I.).

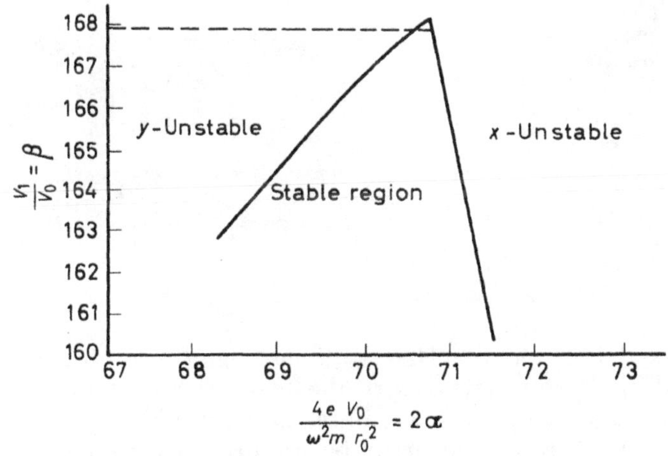

Figure 1.5.5. Stability diagram

Typical stable and unstable trajectories of the ions are shown in *Figure 1.5.4* (a PACE analogue computer was used to obtain these solutions of the equations of motion). A stability diagram has been developed at EAI and is shown in *Figure 1.5.5.*

If the parameter $\beta = V_1/V_0$ is constant at a value of approximately 0·168, there is a very small range of values of the parameter α which will result in stable solutions. Physically this implies that only ions within a very small range of *e/m* ratios will have trajectories which will allow them to traverse the quadrupole structure. Ions with e/m outside this range will have trajectories which result in collisions with the poles.

$$\alpha = (2eV_0)/(mr_0^2\omega^2) \quad \ldots (1.5.9)$$

$$\theta = \omega t \quad \ldots (1.5.10)$$

1.5.3. ION SOURCES

The design of the quadrupole mass filter allows a variety of ion sources to be fitted for special purposes. Some examples of these are shown in the following figures, i.e. axial beam and cross beam ion sources in *Figures 1.5.6(a)* and *(b)*, respectively, and a thermal source in *Figure 1.5.7*. Also being developed are a field ion source and a spark ion source.

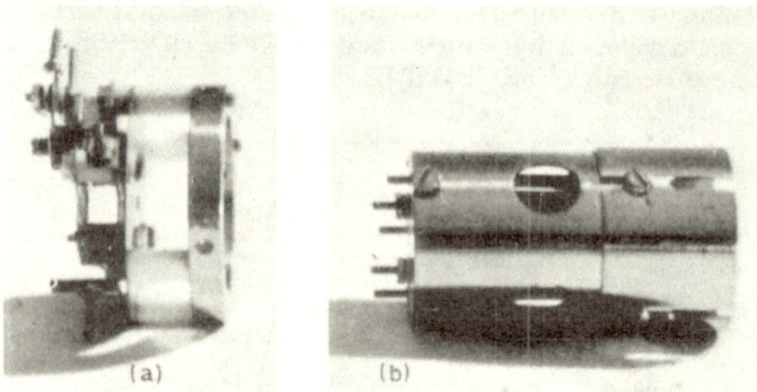

Figure 1.5.6. Ion sources: (a) axial beam; (b) cross beam

The axial beam ionizer has a 'line-of-sight' aperture, through the quadrupole rod system onto the first dynode of the electron multiplier. This makes it very suitable for work with highly reactive molecules which would normally be trapped by intervening metal surfaces such as aperture slits. It is particularly suitable for transient molecules and free radicals, provided their life times are not shorter than the response time of the quadrupole system. Willhoft[4] has recently obtained some interesting results under these conditions and reports further in paper 3.2.

The difficulty with the field ion source is the focusing of the ions once they are formed, since they have a very wide angular spread and even overcome the normally wide acceptance angle of the quadrupole mass filter. Because the efficiency of the field ion source depends on the number of ions accepted, it is necessary to have some prefocusing arrangement which effectively increases the aperture of the quadrupole mass filter. The conventional magnetic deflection instrument has a disadvantage because of its narrow entrance slits[5]. Attempts are being made to overcome this problem by having an initial focusing lens which permits apertures of approximately 1·5 in. diameter, in fact the emitter of the source can be enclosed by the lens and thus a very high collection efficiency

Figure 1.5.7. Thermal ion mounted on quadrupole m.s. head

can be obtained. If this development is successful the applications are numerous and cover such areas as structural analysis, synthesis and process control. One problem outside the quadrupole mass filter, but concerning the actual field ion source, is the stability of the emitter over a period of time, that is, its working life. Currently this seems to be very unstable but is not without the possibility of a solution.

1.5.4. LOW ENERGY ELECTRON DIFFRACTION

One of the most efficient applications of the quadrupole mass spectrometer is in low energy electron diffraction (L.E.E.D.) systems, where magnetic fields cannot be tolerated. L.E.E.D. is now a recognized powerful tool in the study of crystal structures, surfaces and reactions at surfaces. However, the information obtained with these systems was almost impossible to interpret without some fundamental identification of the components present. This could only be provided by a mass spectrometer, and, since the available, suitable mass spectrometers all relied on magnetic deflection, L.E.E.D. systems could not be utilized to their full capabilities—a most frustrating situation. The quadrupole mass spectrometer seems to have provided the answer. (See *Figure 1.5.8.*)

Figure 1.5.8. Quadrupole m.s. on low energy electron diffraction system

The energy of the electron beam is usually in the range 5–500 eV, therefore penetration of the surface is limited to the first few atomic layers. Any contamination of the surface must seriously affect the resultant diffraction pattern. A typical example which demonstrates how this problem may be overcome is the work performed by Tharp, using a quadrupole mass spectrometer on his L.E.E.D. system at the Georgia Institute of Technology[6].

A contaminated sample of tungsten was to be examined. Heating alone caused C^{12} to diffuse to the surface and this was not identified in the L.E.E.D. pattern. The tungsten was flash-heated to 2000°C and the increase of C^{12} evolution was observed by a change in system pressure from 10^{-11} torr to 10^{-10} torr during degassing (masses other than C^{12} decreased). Oxygen was admitted, which combined with the C^{12} to form carbon dioxide. The sample was then heated to 1000°C, in an atmosphere of oxygen at 10^{-8} torr, followed by flash heating to 1700°C. During the latter heating, maximum desorption of carbon monoxide (mass 28) was observed without any great increase in C^{12}. This experiment was performed efficiently with the aid of the quadrupole mass spectrometer.

Other surfaces, such as chromium and titanium, which cannot be heated as can tungsten, can be cleaned by ionic bombardment in an inert gas such as argon. Here the quadrupole mass spectrometer can be used to determine the purity of the ion beam and to sense the degree of contamination of the specimen.

A novel feature of this L.E.E.D. system is the automatic gas environment control system. The quadrupole mass spectrometer has been modified to monitor four gaseous components simultaneously. This is performed by switching the 0–15 V input to the r.f./d.c. generator at preselected points, in sequence, on a variable time interval. For instance, masses 16, 18, 28 and 44, can be repeatedly sampled, in sequence with a 10 ms interval, thus giving a quasi-continuous recording of the concentration of each component with respect to time. A further stage is to couple this system to servo drives which operate the gas inlet valves, enabling any gas environment to be obtained completely automatically, from the four components. This sophisticated gas control system can be used for studying surface reactions in the presence of a predetermined atmosphere. This facility is now becoming standard equipment on most quadrupole mass spectrometers.

1.5.5. MOLECULAR BEAM SYSTEMS—ION-MOLECULE REACTIONS

In molecular beam experiments, particularly in low energy beam work, the quadrupole mass spectrometer has found at least two major applications; one as a magnetic field-free detector and another as a source of ions of a particular species. As a detector in these experiments the quadrupole mass spectrometer head is light, weighing less than 2 lb. It is therefore easy to mount as a directional detector scanning a reaction zone, such as the scattering centre in a crossed molecular beam experiment, with or without the ion source.

Again, the presence of a magnet, or a magnetic field, would render the application impossible. The use of the quadrupole mass spectrometer as an ion source of a particular species is interesting.

In practice the collector (either a Faraday cup or an electron multiplier), is removed and is replaced by a focusing lens. This focusing lens acts as an ion beam collimating lens. The ions, when they exit from the rods under normal conditions, are spread over a wide solid angle as high as 120°. The ions after being collimated by the post rod lens, pass into a further focusing lens where their velocity is also controlled. The rest of the quadrupole mass spectrometer head remains virtually untouched and is operated normally.

1.5.6. PROCESS CONTROL

It is now being realized by workers with quadrupole mass spectrometers that these instruments have a very large potential market in industrial process control. One area where they have been considered is in on-stream analysis of chloro-hydrocarbon products, where they would replace existing gas chromatographic separation and identification systems. One quadrupole mass spectrometer head would probably be able to perform the work of several g.l.c. systems in a much shorter time. It would be more reliable, very much more compact and more adaptable to the output of the analytical information, in the required form, for control of the plant to which it is attached.

With this application several essential requirements must be met. Some of these can be listed generally, but specific production processes may have their individual needs. In general, the following are required:

(*i*) A suitable bakable inlet system which will provide the necessary pressure drop from the process pressure to that acceptable to the quadrupole mass spectrometer (10^{-4} torr max., 10^{-5} torr working).

(*ii*) A twin pumping system which will handle the pressure drop gas load and enable one system to be on continuous standby in case of maintenance or failure of the on-stream equipment.

(*iii*) Ability to run the quadrupole mass spectrometer head at an elevated temperature, in fact the process temperature if possible, or up to 250°C.

(*iv*) Remote control of the quadrupole mass spectrometer head, i.e. the siting of most of the electronic equipment in a control room which may be well away from the quadrupole mass spectrometer head.

(*v*) High speed scanning and ability to monitor a mass component continuously as a d.c. level.

(*vi*) Ability to monitor up to 10 individual species as a quasi-simultaneous display and possibly at a variable ionization potential.

The quadrupole mass spectrometer can be readily adapted to these requirements and a typical plant is shown schematically in *Figure 1.5.9.*

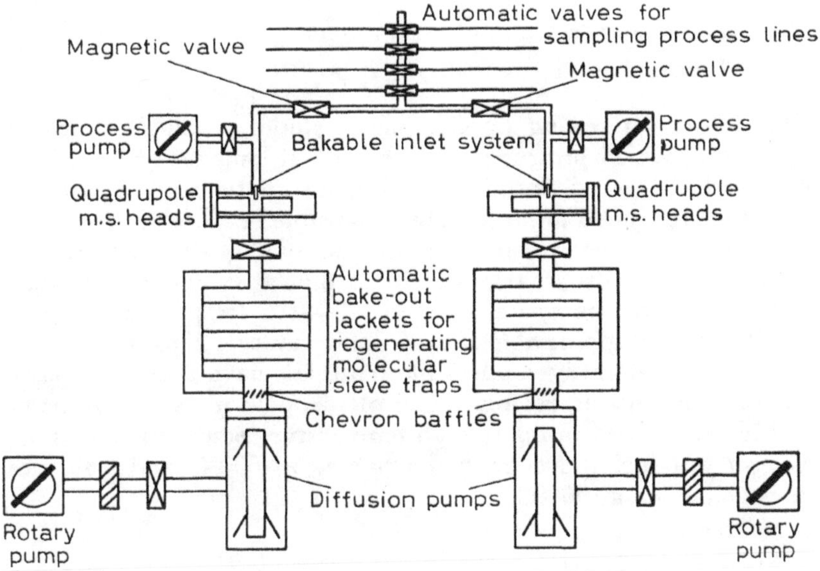

Figure 1.5.9. Twin pumping and analyser for 'on-stream' process control

The analytical information can be fed into a digitizer, then an analogue module, where information from a particular stream can be compared with a calibration pattern for the process and the required control loops operated to make any necessary adjustment to the process. As in the automatically controlled gaseous environment circuit developed for the L.E.E.D. system mentioned earlier in this paper, the degree of automation is arbitrary, which makes the quadrupole mass spectrometer much more versatile in this application.

1.5.7. MASS SPECTROMETER—GAS CHROMATOGRAPH LINK

The quadrupole mass spectrometer has probably more to offer in this field than any other and is being extensively used by many workers. The following are some of the major advantages:

(*i*) High sensitivity.

(*ii*) Low cost.

(*iii*) Large mass range.

(*iv*) High speed scanning.

(*v*) Ability to monitor several selected components simultaneously.

(*vi*) Automatic selection of mass range.

(*vii*) Linear mass separation.

The more interesting applications include an all-glass heated inlet system which protrudes into the actual ionizing region of the ion source, allowing unstable compounds to be ionized without contacting any metal surfaces. This is possible because the geometry of the ion source is very open and no collimating slits are required. A further extension of this system is that the quadrupole mass spectrometer can be operated at an elevated temperature (up to 250°C) without appreciable loss of performance. This means that non-volatile and easily condensible materials have a better chance to enter the ionizing region. The quadrupole mass spectrometer head has been used in many high temperature environments. One has been producing spectra while baking at 600°C and a further one is being run at 900°C.

1.5.8. FAST CHEMICAL REACTIONS

The quadrupole mass spectrometer has found some of its earliest applications in the field of fast chemical reactions, particularly combustion, photolysis, and ions in shock waves. The high speed response and the compactness of the quadrupole mass spectrometer head are particularly important. One typical application has been the analysis of motor car exhaust gases, where the composition of the exhaust gases of efficient and inefficient engines has been compared directly, while they have been running under test conditions. The use of the modification, which enables several gaseous components to be monitored simultaneously, was specially important in this application. The day when motor cars are tuned with a mass spectrometer instead of electronically may not be so far away.

Finally, *Table 1.5.1* shows, as far as I am aware, the existing commercial manufacturers of quadrupole mass spectrometers and some of their relative performance figures. It would seem that they are very good value for money when compared with other types of mass spectrometer.

Table 1.5.1. Comparison of Available Quadrupole Mass Spectrometers

Manufacturer	*Mass range*	*Resolution $m/\Delta m$–10% valley*	*Sensitivity*	*Highest scan speed*
EAI centronic				
Quad 300	1–500 u	500	10 A/torr N_2 *pp* (10^{-15} torr N_2)	1–50 u/50 ms
Quad 250	1–500 u	500	10–100 A/torr N_2 *pp* (10^{-15} torr N_2)	1–50 u/50 ms
Quad 150	1–150 u 1–250 u	150 250	5 A/torr N_2 *pp* (10^{-14} torr N_2)	1–150 u/130 ms
MAT				
Amp 3	1–200 u	150	*pp* (4×10^{-14} torr N_2)	1–120 u/100 ms
Varian associates				
QRGA	1–250 u	50	*pp* (5×10^{-11} torr N_2)	1–50 u/50 ms
Balzers aktiengesell-schaft QMG 101	1–400 u	100	*pp* (10^{-13} torr N_2)	1–100 u/100 ms

ACKNOWLEDGEMENT

The author acknowledges the kind permission of the Directors of 20th Century Electronics Ltd. to present this paper, and the valuable assistance given by his colleague Mr. C. J. Bargery.

REFERENCES

1 McLachlan, N., *Theory and Applications of Mathieu Functions*, Oxford University Press, London, 1947
2 Paul, W., Reinhard, H. and Von Zahn, Y. *Z. Phys.* 152 (1958) 143
3 Rice, D., *Quadrupole Mass Spectrometer*, Interim Report No. 2, Stanford Research Institute, 1962
4 Willhoft, E. M. A. and Robertson, A. J. B. *Chem. Commun.* (1967) 385
5 Beckey, H. D., *Mass Spectrometry*, R. I. Reed (Ed.), Academic Press, New York, 1965
6 Tharp, L. N., *L.E.E.D. The Answer to your Surface Problems?*, Applications Bulletin, Electronic Associates Inc., Palo Alto

SESSION 2

PHYSICAL AND CHROMATOGRAPHIC APPLICATIONS

2.1

A FAST-SCAN MASS SPECTROMETER FOR RESIDUAL GAS ANALYSIS AND THE EXAMINATION OF EFFLUENTS FROM GAS CHROMATOGRAPHY COLUMNS

S. Evans and R. D. Collins

2.1.1. INTRODUCTION

THE importance of partial pressure analysis in vacuum research has been recognized for many years and mass spectrometers are now beginning to be used on production plant for process monitoring and control. Instruments developed for residual gas analysis include several 180° magnetic deflection types, the Omegatron and some small r.f. tubes. In general these have a minimum partial pressure sensitivity of about 10^{-11} torr and a scan rate of the order of 5–10 s per mass peak, giving a scan time over 50 mass units of from 4 min to 8 min. The r.f. tubes, although providing an inherently fast scan rate, generally suffered from low sensitivity and very poor resolving power.

For most applications such instruments are more than adequate, but with the advance of vacuum technology, a demand has arisen for instruments capable of recording much lower partial pressures. For total pressures of 10^{-10} torr, for example, the minimum detectable partial pressures should be of the order of 10^{-13} torr. Similarly, for gas analysis in dynamic systems operating at total pressures in excess of 10^{-8} torr, for example, the study of outgassing products during vacuum tube processing, gettering and all systems in which sorption and desorption are taking place, much faster scan rates are required to follow rapid changes in gas composition. In the field of chemical research, the gas chromatographers have discovered the mass spectrometer, and demand instruments which will analyse minute fragments of chemical mixtures, which can only be present in the mass spectrometer for a few seconds. In many cases, a relatively simple instrument of limited mass range and resolution is adequate for this purpose.

The minimum detectable partial pressure of an instrument and the maximum usable scan speed are both dependent on the parameters of the d.c. amplifier used to detect the ion beam. The fundamental limitations are thus the thermal noise of the input resistance and the fluctuation noise of the amplifier itself. This depends not only on the value of the feedback resistance, but also on the bandwidth of the amplifier which, in turn, depends on the required response time, i.e. on the time allowed for scanning a single peak. It follows that, if any significant improvement in partial pressure sensitivity or scan rate is to be achieved, the current available at the amplifier input must be increased. The simplest way in which this can be done is to use a particle multiplier detector, which is, in effect, a fast response preamplifier with a current gain of, say, 10^5, and a noise level of the order of 10^{-19} A, referred to the input. With this arrangement noise and response-time limitations are reduced by several decades, the main limitation now being statistical noise caused by fluctuations in the number of ions arriving at the first dynode of the multiplier[1, 2].

A small sector field mass spectrometer, specifically designed for residual gas analysis and incorporating a particle multiplier, has been developed with a high intrinsic sensitivity even under fast scanning conditions. This instrument has also proved suitable for use with gas chromatography equipment and has been used for this purpose for the past 2 years.

The mass spectrometer head (*Figure 2.1.1*) was a 3 in. radius, 60° sector field instrument of all stainless steel construction, and bakable to 350°C. Aluminium wire seals were used throughout. The analyser body consisted of a machined central section which was used as an accurate assembly jig for the rest of the instrument. This section carried the source and collector flight tubes, a central pumping port, and magnet location. The source and collector flanges were welded to the flight tubes and machined to accept the ion source and collector subassemblies. The ion source was mounted proud of the source flange, so that for residual gas work, the mass spectrometer could be used with a nude source. For other applications, including gas chromatography, the source was covered by a stainless steel 'top hat' and the tube pumped via the central port. Separate pumping of the source and analyser sections was also possible. Permanent magnets were generally used, a choice of four magnets giving a total mass coverage of 1–400 u.

Scanning was by variation of the ion accelerator potential, and two independent scanning systems were incorporated in the ion accelerator unit. One of these used a motorized helipot giving a reciprocal law output over a voltage range of 5 kV to 500 V, at rates of from 0·9 min to 900 min per scan. This gave linear mass

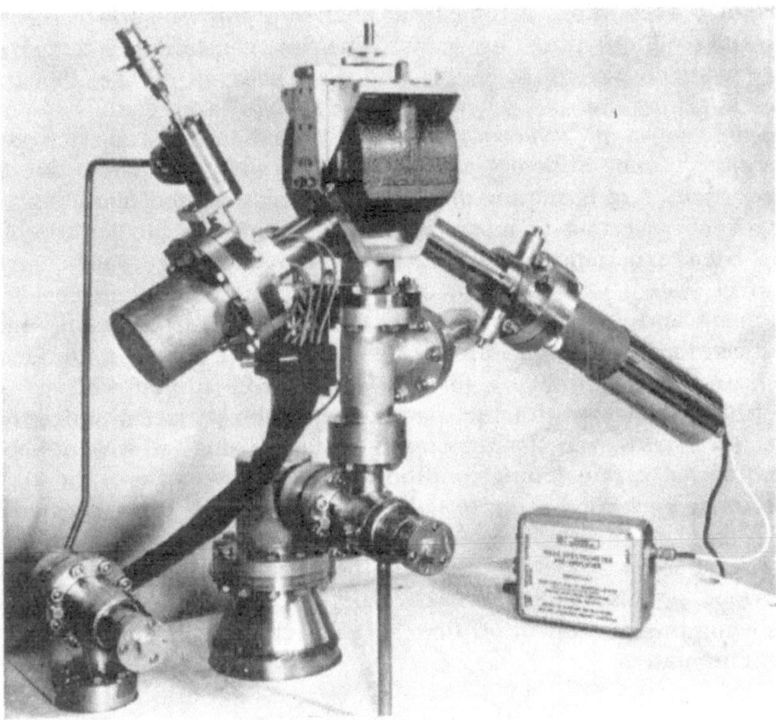

Figure 2.1.1. 60° *Mass spectrometer head with* 6100 G *magnet.* (*From* W. Steckelmacher, S. Evans and R. Collins[3], *by courtesy of* Nuovo cimento)

separation and covered a mass ratio of 10:1—i.e. a 6100 G magnet covered the range 26–260 u. The second scan system was an exponential decay electronic scan giving scan rates from 0·1 s to 10 s per mass range, in this case from 5 kV to 200 V. The exponential law was chosen because this gives a constant peak width over the mass range, assuming constant resolving power, and hence constant peak rise time. Mass coverage in this case was over a 25:1 ratio, thus the 6100 G magnet covered a theoretical mass range of 26–650 u. In practice, the upper mass limit was about 400 u, due to deterioration of resolution and sensitivity above this value. Residual gas peaks have, however, been observed up to about 510 u.

The ion source was of the electron bombardment type and was constructed as a subassembly on a small demountable flange. Space-charge control of the electron beam was preferred to the more conventional temperature control, to reduce the effect of emission fluctuations arising from changing emitter surface conditions during individual fast scans. The time constant of a temperature control system is necessarily slow because of the thermal inertia of the filament, whereas for the space-charge control system the time constant of the control loop is limited only by the electronic circuitry. Some difficulty was experienced with this system due to the formation of insulating deposits when using coated filaments[3–5], but since reverting to pure rhenium filaments, no further trouble has been experienced. Another advantage of the space-charge control system was that the control grid interposed between the filament and ion box helped to focus the electron beam, and reduced the electron current lost to the ion box from approximately 150 μA to less than 20 μA, for a 50 μA electron trap current.

When operating with the space-charge control system and using the 0·1 s scan over the maximum voltage swing, it was noticed that, under certain tuning conditions, the first few peaks of the mass spectra were severely suppressed. This was found to be associated with the fast scan flyback, which was rising from 200 V to 5 kV at a rate of some 5×10^5 V/s. It is thought that this was inducing spurious signals in the source cables, which were momentarily throwing the instrument off tune. At slower flyback rates the effect was eliminated.

2.1.2. PERFORMANCE

The sensitivity of the instrument was approximately 0·4 per torr at an electron beam current of 50 μA. This corresponds to an ion beam current of 2×10^{-15} A for a partial pressure of 10^{-10} torr, nitrogen equivalent. The resolution obtained with the standard defining slits was approximately 150 with a 5% valley. Narrower slits, which resulted in complete separation (i.e. to baseline) of the mercury isotopes with little loss of sensitivity, have been tried.

2.1.3. THE ION DETECTOR

The ion collector subassembly contained a double-ended Faraday cage collector with an interposed 50% transmission grid. This grid intercepted half the ion beam, the remainder continuing to the first dynode of the particle multiplier, which was mounted on a separate flange. The multiplier gain could be readily checked by comparing the ion current at the grid and the electron current output from the multiplier. The grid collector was also used to measure ion beams too large to be handled by the multiplier.

The amplifier was designed to operate with both ion or electron current inputs and incorporated three switched feedback resistors of $10^7\ \Omega$, $10^9\ \Omega$ and $10^{11}\ \Omega$ with a voltage sensitivity for each resistor of 30 V to 30 mV full scale deflection in seven ranges. Response time for the $10^7\ \Omega$ range was 50 μs corresponding to the fastest scan rate of 200 μs per mass peak. At this scan rate slight peak clipping is inevitable. This specification was chosen to give partial pressure sensitivities of approximately 10^{-10} torr, 10^{-12} torr and 10^{-14} torr for 10% deflection on the most sensitive range, for each feedback resistor, and assuming an average multiplier gain of 10^5.

The amplifier included outputs for an oscilloscope, a potentiometric recorder, three ultra-violet recorder galvanometers and meter display.

2.1.4. NOISE AND SCAN RATE LIMITATIONS USING PARTICLE MULTIPLIERS

When considering the measurement of very small currents, it is necessary to think in terms of the rate of flow of charge carriers, 1 A being equivalent to $6{\cdot}25 \times 10^{18}$ electrons, or singly charged ions, per second. Thus an ion beam of 2×10^{-15} A would consist of 12,500 ions per second. The percentage standard deviation (s.d.) of this current is given by:

$$\text{s.d. } (\%) = 100\ (nt)^{-1/2}$$

where n is the number of particles arriving per second and t is the time constant of the measuring system. For the current quoted and a time constant of 50×10^{-6} s the percentage s.d. would be 125%. This figure sounds excessive, but for qualitative work, a detectable signal is obtained as the noise signal only occurs on a spectrum where a peak is present, the background being unaffected and stable. These figures would apply to the oscilloscope and u.v. recorder outputs only, as the meter and potentiometric recorder would have inherent response times of about 0·5 s and would display an average output. With a 0·5 s response time, the percentage s.d. for the same signal would be 5%, but the scan rate would need to be reduced to about 5 s per peak to avoid peak clipping.

A further factor which must be taken into account is the number of ions sampled from the ion beam during the scanning time of a single peak. The peak obtained from an ion beam of 2×10^{-15} A which is scanned in 200 μs would contain an average of $12{,}500 \times 200 \times 10^{-6}$, which equals 2·5 ions.

The gain of the particle multiplier is an important factor in the above calculations, which all apply to a 10% deflection on the most sensitive voltage range of the amplifier and for a gain of 10^5 with a $10^7\,\Omega$ feedback resistor. If the gain was 10^6, the same deflection would correspond to an ion beam of one-tenth the intensity, i.e. 0·25 ions in 200 μs or more—practically one ion every four scans. The multiplier e.h.t. supply must consequently be adjustable, to enable the gain to be matched to the amplifier and the required current and scan rate.

The figures used in the calculations above are those appertaining to the mass spectrometer described, but the very simple principles involved apply to all mass spectrometers and similar devices. Sensitivity and scan rate, and to some extent resolving power, are interdependent factors and must be treated as such when assessing the performance of a particular instrument.

2.1.5. APPLICATION TO RESIDUAL GAS ANALYSIS

Several of these instruments are at present in use in our laboratories, investigating different aspects of vacuum technology. One instrument is being calibrated as a substandard partial pressure gauge in the mass range 1–50 u. In the course of this work, investigations were carried out on the effect of mechanical and electrical tolerances on the partial pressure sensitivity, e.g. the position and orientation of the source magnet, the main analyser magnet and filament, and the effect of small changes in source electrode potentials. It is hoped that the final calibration of this instrument will be carried out using the gauge calibration system designed by Fletcher *et al.*[6], which provides a partial pressure accuracy for permanent gases of ± 3%, down to 10^{-8} torr.

A second instrument is being used for residual gas investigations at u.h.v. using conventional diffusion pumped systems, and the influence of baking procedure and pretreatment of components on the residual gas spectra[7]. An interesting aspect of this work was the effect of outgassing the ionization gauge grid on the residual gases. A portion of a repetitive fast scan taken, using a limited mass range during grid outgassing, is shown in *Figure 2.1.2*. The portion shown represents conditions some 4–5 s after commencing outgassing at approximately 50 W. The spectrum before outgassing was a simple one consisting primarily of masses 28 and 44 with a small peak at mass 32. The outgassing reached a maximum within 1 s and is clearly falling rapidly over the few seconds covered in the illustration. During the first 2 s or 3 s, a number of peaks were apparent at high mass numbers, including those corresponding to mercury isotopes.

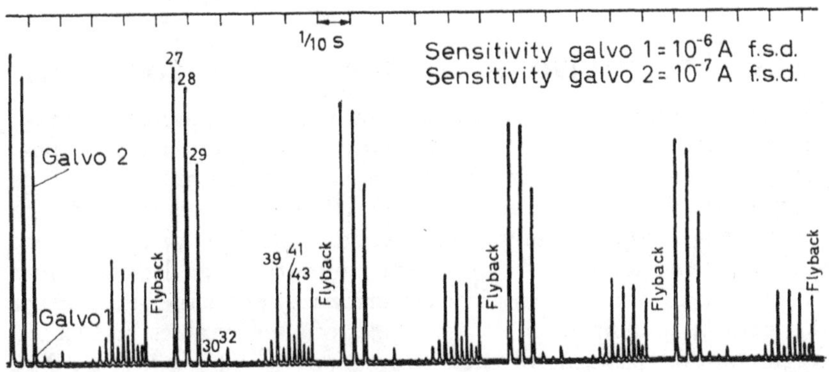

Figure 2.1.2. Portion of repetitive scan taken during outgassing cycle of ion gauge (range 27–45 u). (*From* W. Steckelmacher, S. Evans and R. Collins[3], *by courtesy of* Nuovo cimento)

Redhead[8] has described the so-called desorption spectrometer in which a tungsten filament is allowed to sorb residual gas for a known period. The temperature of the filament is then raised slowly and an ionization gauge used to measure the pressure changes caused by the desorption of the gases previously adsorbed by the filament. Desorption of the components of the residual gas takes place at different temperatures, depending on the heat of desorption of the component. Partial pressures normally well below the detection limit of the gauge can be detected in this way.

From our observations it would seem that this effect can be used to extend the usefulness of the fast-scanning mass spectrometer. Partial pressure sensitivity enhancement could be achieved by sorption over a period, as with the desorption spectrometer, using an ion gauge filament or grid, or a specially provided filament. The sorbed gas could be desorbed rapidly to give a quick assessment of the conditions of the vacuum system, or slowly to give separate desorption and analysis of individual components and hence, easier interpretation of mass spectra.

A third instrument is being used to study problems associated with the use and performance of Penning and radial field pumps, and some aspects of this work have been described previously by Fulker, Holland and Laurenson[9, 10]. They used the instrument to show that the residual gas atmosphere of a high vacuum system using these pumps is dependent on the method used to pre-exhaust the system before starting ion-pumping. Pre-exhaustion by liquid nitrogen cooled sorption pumps containing molecular sieves resulted in higher partial pressures of helium and neon than when a rotary pump was used together with an activated alumina trap.

2.1.6. APPLICATIONS TO GAS CHROMATOGRAPHY

Gas chromatography is a technique by which a complex chemical mixture can be separated into its individual components. These components are eluted from the chromatograph column over periods ranging from a few minutes to several hours, the elution of each component occupying only a few seconds. Qualitative information and the identification of components must then be deduced from retention time data and must be confirmed by other means. Passing the column eluent into a mass spectrometer enables individual fragmentation patterns of each component to be produced. The most convenient way in which this can be done is to use a fast-scanning mass spectrometer, which is capable of producing several mass spectra from each component as it appears from the column. Components which are imperfectly separated by the column can often be identified in this way by examination of successive scans, showing the rise of the fragmentation pattern from one component as that of the other component falls.

The requirements for a mass spectrometer suitable for this application are, in many ways, similar to those of a u.h.v. residual gas instrument. The instrument must be bakable, if it is required to analyse any but the major components of the injected sample, in order to eliminate background interference. This implies the use of a completely inorganic system with metal seals throughout. The ion source and inlet system must be capable of operating at the same temperature as the chromatograph in order to prevent condensation and hence loss of sample. The upper limit here is the bulk resistivity, at elevated temperatures, of the insulating materials used and materials used for external plugs, sockets and cables. This limit is usually of the order of 250–300°C.

The partial pressure sensitivity must be as high as possible, and the scanning rate not greater than 1 or 2 s over the mass range. In most cases, the mass range need not be very extensive, with an upper limit of 400–500 u and a lower limit of 26 u. For a wide range of applications an upper limit of 300 u is probably adequate. McLafferty[11] has examined spectra from 4036 compounds and tabulated the masses of the three most abundant peaks from each spectra. Only about a dozen peaks fall above 260 u and about 100 below 26 u. It is not suggested that the three most abundant peaks are necessarily sufficient to identify any component, but certainly a considerable amount of information can be obtained from spectra below 260 u. Low resolution also has advantages in that the consequent low ratio of peak separation to peak width enables faster scanning to be used.

The 60° instrument described fulfils most of these requirements, and a fourth instrument in our laboratories is being used in combination with a gas chromatograph (Philips M1400). The equipment uses an 18 ft long, $\frac{1}{8}$ in. bore, stainless steel column, packed with 10% Apiezon L on Cellite 100–110 mesh.

Various forms of the combined instrument have been successfully applied to a range of problems, from the analysis of flavour volatiles in such foods as potatoes, chicken and bananas, to the investigation of degradation products of rotary pump oils.

2.1.7. DESIGN OF A COMBINED GAS CHROMATOGRAPHY–MASS SPECTROMETER SYSTEM

Most of the problems involved in the design of a combined gas chromatography–mass spectrometer system are concerned with the interconnecting stages. The flow rate of carrier gas through the chromatograph can vary between 1 ml/min and 50 ml/min, depending upon the type of column chosen. A mass spectrometer with an operating source pressure of 10^{-5} torr and source pumping speed of 10 l/s would accept only 10^{-4} l torr/s or $0{\cdot}8 \times 10^{-2}$ ml/min, i.e. from one part in 100 to one part in 5000 of the total flow. Differential pumping of the source and analyser regions would enable higher flows to be accepted.

The flow splitter chamber shown in *Figure 2.1.3* was designed to accept a flow rate from atmospheric pressure of 1 ml/min. The column flow was allowed to vent to atmosphere after flowing past the splitter inlet, thus allowing the splitter to accept at the required rate, independent of column flow. The splitter inlet consisted of a short length of 0·010 in. bore stainless steel capillary, the outlet of which entered a small vacuum enclosure. The molecular beam from the capillary was directed at a small glass pinhole leak which formed the mass spectrometer inlet. A replaceable conductance was mounted in a side arm of the vacuum enclosure, the bore of which was adjusted to give a conductance of 1000 times that of the glass leak.

The splitter was mounted on the ion source sample inlet flange, and the complete ion source and splitter assembly could be operated at temperatures up to 250°C. When using a long capillary column it was possible to identify components of about 5 μg using this system.

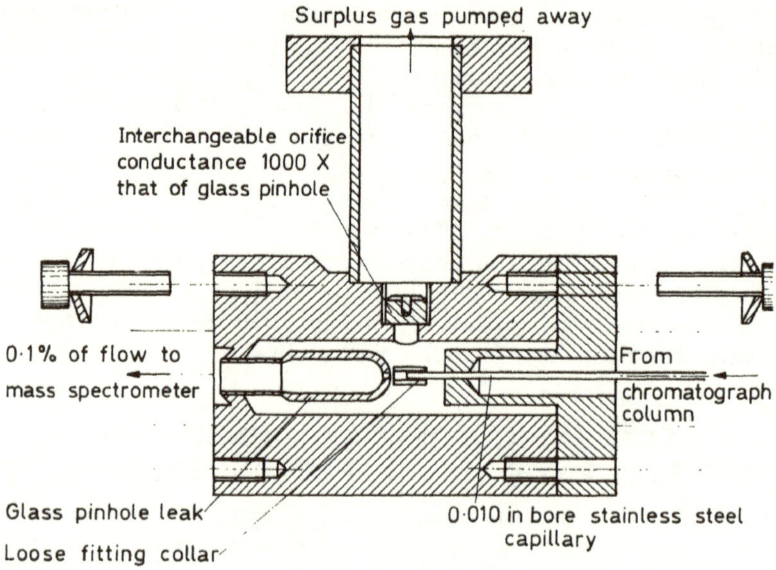

Figure 2.1.3. Single-stage flow splitter for coupling a gas chromatograph to a mass spectrometer

2.1.8. PREFERENTIAL REMOVAL OF HELIUM

The total pressure in the mass spectrometer, when used with a gas chromatograph, is controlled entirely by the quantity of carrier gas allowed to enter it. Helium is usually preferred as a carrier gas because of its inert nature, low molecular weight, which leads to minimum interference from the carrier gas peak with the fragmentation patterns of the sample being analysed, and relatively high pumping rate. The minimum sample size that can be detected is dependent upon the effective concentration of the sample in the carrier gas entering the mass spectrometer, and, if this concentration can be increased, the sample size can be decreased.

Several methods have been suggested for increasing sample concentration, most of which depend upon the preferential removal of helium from the gas stream in the inlet splitter. One method used is the separation jet principle originally developed by Becker[12] for the separation of isotopes, and since developed by Ryhage[13] as a chromatograph inlet system. In this device a stream of gas consisting of two or more components of different molecular weights is passed through a capillary jet into a low pressure chamber and directed at a second capillary jet positioned in line with, and a few thou. in front of, the orifice of the first jet. The lighter component of the gas mixture tends to diffuse sideways by pressure diffusion, whereas the heavier components tend to continue as a molecular beam into the second capillary.

The splitter described above could readily be converted into a single-stage separation jet by accurate aligment of the 0·010 in. capillary with the glass pinhole leak and by reduction of the orifice

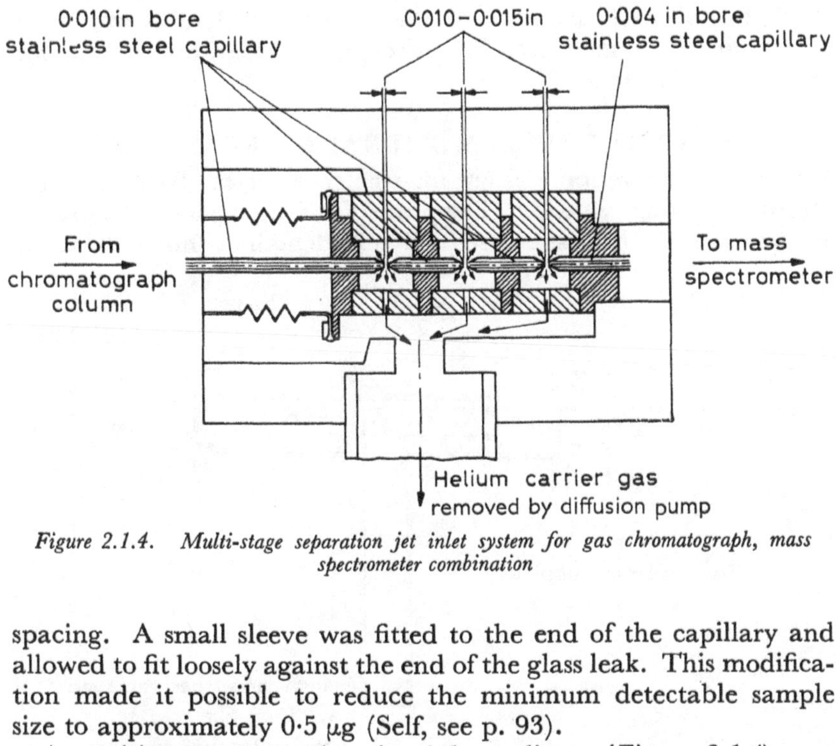

Figure 2.1.4. Multi-stage separation jet inlet system for gas chromatograph, mass spectrometer combination

spacing. A small sleeve was fitted to the end of the capillary and allowed to fit loosely against the end of the glass leak. This modification made it possible to reduce the minimum detectable sample size to approximately 0·5 μg (Self, see p. 93).

A multi-stage separation jet inlet splitter (*Figure 2.1.4*) was manufactured and tested. This used three separation jet stages, each consisting of a small chamber with a 0·010 in bore inlet capillary and 0·010 in. bore outlet capillary, apart from the final stage where the outlet capillary was 0·004 in. in diameter. Each chamber was pumped through a 0·020 in. diameter by $\frac{1}{8}$ in. long hole into an outer vacuum jacket which, in turn, was pumped by a diffusion pump at approximately 10 l/s pumping speed. The pressure was arranged to drop by a decade in each chamber so that the first chamber operated at 10 torr, the second at 1 torr, the third at 10^{-1} torr and the outer jacket at 10^{-2} torr. The performance of the separator was assessed by injecting a known quantity of benzene vapour into the chromatograph and measuring the height of the benzene 78 peak and the helium peak under various conditions of pressure, temperature and flow rate. The peak heights were compared to those obtained under identical conditions, with the unmodified flow splitter described previously. The maximum improvement in the benzene peak height obtained was approximately 20 to 1 and performance was sensitive to both pressure and flow. No doubt this performance could have been considerably improved by careful investigation of the effect of jet separation and other mechanical tolerances.

2.1.9. PERMEABLE MEMBRANE SEPARATORS

A permeable membrane separator was constructed, based on that described by Lipsky *et al.*[14] The principle of this device is illustrated in *Figure 2.1.5.* It consists essentially of a length of thin wall, small

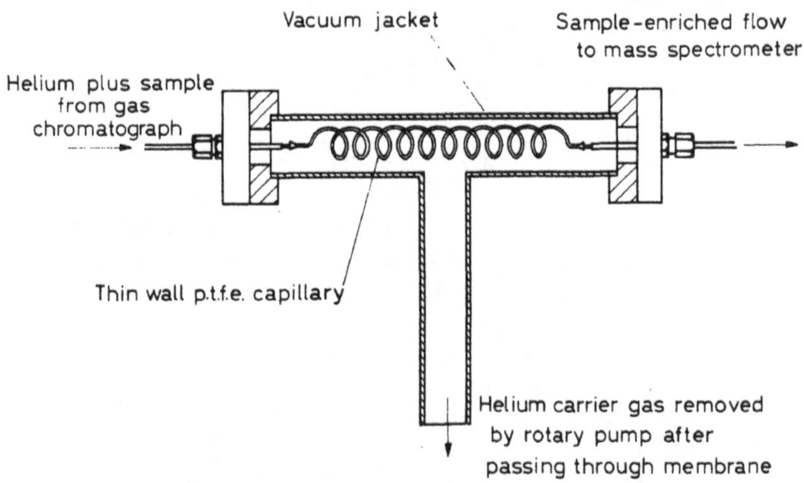

Figure 2.1.5. Permeable membrane helium separator for use with gas chromatograph, mass spectrometer combination

bore, polytetrafluoroethylene (p.t.f.e.) tube, the ends of which were coupled to short lengths of stainless steel capillary brazed into flanges. This assembly was mounted in a stainless steel vacuum jacket which was pumped through a ½ in. internal diameter stainless steel tube by a single-stage rotary pump (Edwards model ES35). The separator was enclosed by a small oven and maintained at a constant temperature of the order of 275°C. The total length of p.t.f.e. used was about 7 ft. The inlet to the separator was through a length of 0·010 in. bore stainless steel capillary, arranged to give a flow rate through the separator when at the maximum operating temperature, of 2 ml/min. The variation of helium flow with temperature is shown in *Figure 2.1.6.* A second length of 0·010 in. bore capillary was used to connect the separator to the mass spectrometer, suitably constricted to give a flow rate to the mass spectrometer of approximately 10^{-2} ml/min.

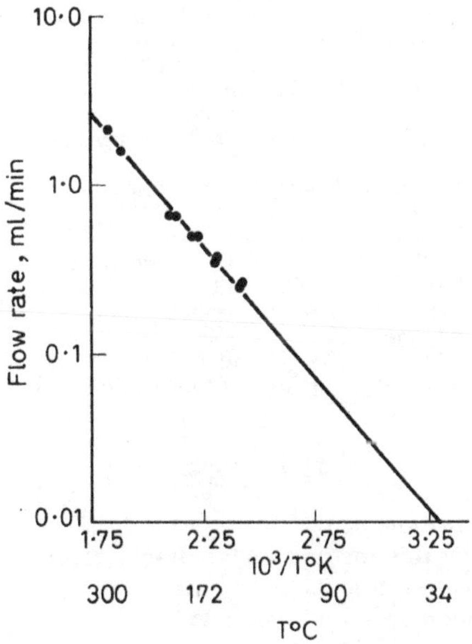

Figure 2.1.6. Flow rate of helium through permeable membrane separator as a function of temperature

With this separator, most of the helium carrier gas is removed by permeation through the wall of the p.t.f.e. tube, therefore most of the 10^{-2} ml/min flow to the mass spectrometer was sample material. Assuming no loss of sample through the p.t.f.e., the amount of sample entering the mass spectrometer would be increased roughly in the ratio of the flow rates into and out of the separator, i.e.

200 to 1. Satisfactory spectra have been obtained from a few nanogrammes of sample (diethylether) using this inlet system.

The authors acknowledge the work of Mr. J. English (electronic design), Mr. J. F. Watts (mechanical design), Mr. T. Chapman (manufacture of ion sources, etc.) and Messrs. M. Parrish and A. Houghton, who, as Diploma of Technology students, carried out most of the work on the gas chromatograph inlet systems described.

REFERENCES

1 BRUNEE, C., JENCKEL, L. and KRONENBERGER, K. *Z. analyt. Chem.* 189 (1962) 50
2 BEYNON, J. H., *Mass Spectrometry*, Elsevier, New York, 1960
3 STECKELMACHER, W., EVANS, S. and COLLINS, R. D. Supplto nuovo Cim N.1, Serie 1, 5 (1967) 224
4 WINN, E. B. and NIER, A. O. *Rev. scient. Instrum.* (1949) 773
5 JEFFRIES, D. K. *J. scient. Instrum.* 44 (1967) 587
6 FLETCHER, B. and WATTS, J. F. *Vacuum* 17 (1967) In press
7 EVANS, S. and COLLINS, R. D. *Vide* In press
8 REDHEAD, P. A., 'The Desorption Spectrometer as an Analytical Tool in U.H.V. Investigations', *6th Natn. Symp. Vacuum Technol. Trans.*, Pergamon, London, 1959
9 HOLLAND, L., FULKER, M. J. and LAURENSON, L. Supplto nuovo Cim N. 1, Serie 1, 5 (1967) 242
10 FULKER, M. J., HOLLAND, L. and LAURENSON, L. *Vide* In press
11 MCLAFFERTY, F. W. 'Mass Spectral Correlations', *Am. Chem. Soc., Adv. Chem. Ser.* 40, Washington D.C., 1963
12 BECKER, E. W., *Separation of Isotopes*, Newnes, London, 1961
13 RYHAGE, R. *Analyt. Chem.* 36 (1964) 759
14 LIPSKY, S. R., MCMURRAY, W. J. and HORVARTH, C. G., *Gas Chromatography*, A. B. LITTLEWOOD (Ed.), Institute of Petroleum, London, 1967

DISCUSSION

J. C. HALL (WILD-BARFIELD)
Is it possible to use this method for analysing gases extracted from metal specimens by vacuum fusion, particularly the quantities of hydrogen, nitrogen, carbon monoxide and methane?

S. EVANS, in reply. You should be able to do this with a small vacuum fur nace attached near to the input of the instrument, the main difficulty will be the establishment of a low background mass spectrum before melting the sample and assessing the quantity, together with the difficulty of separating nitrogen and carbon monoxide. It is doubtful if quantitative information to the accuracy required could be readily obtained.

D. S. ALLAM (UNIVERSITY OF SURREY)
Could you expand on your comment that insulating deposits were obtained from filaments using lanthanum hexaboride on rhenium?

S. EVANS, in reply. This arose from overheating the filament, at that time under space-charge control conditions, particularly during the activation period. We suspect that this resulted in the deposition of material on the control electrode to produce a non-conducting coating. In more recent experiments, the emitter was activated under temperature controlled conditions and hence at a lower temperature, and no further difficulty was experienced.

Note. Since making these comments, Faureau and Koenig[1] have suggested that lanthanum hexaboride, as supplied from some manufacturers, is impure and that the impurities are most likely to be either free boron or boric oxide, of which the boric oxide is very volatile. Our own observations would be more satisfactorily explained by the presence of these impurities than by a basic property of the lanthanum hexaboride itself.

Faureau and Koenig were interested in reactions taking place with the base metal (rhenium) and suggested a purification procedure which effectively eliminated the impurities and, hence, the base metal reaction.

[1] Faureau, L. J. and Koenig, D. F. *Rev. scient. Instrum.* 38 (1967) 841

2.2

FLAVOUR RESEARCH WITH AN INEXPENSIVE FAST-SCANNING MASS SPECTROMETER

R. Self

2.2.1. INTRODUCTION

GAS chromatography (g.c.) has been used in conjunction with a mass spectrometer (m.s.) for over a decade[1, 2] but, until the recent advent of fast-scanning mass spectrometry (f.s.m.s.), an efficient direct-coupled combination (g.c.–m.s.), utilizing the attributes of both techniques, was not possible. Separately, both g.c. and m.s. provide information which can be used to characterize a volatile substance; the mass spectral data will be most accurate when obtained from a slow-scan spectrum (i.e. no 'peak clipping' caused by random errors at the collector), and the gas chromatographic data will be most reliable when the column flow is not affected by the conditions necessary for introduction of the sample into the mass spectrometer. Unfortunately it is common to have only sub-microgramme quantities of sample available in the extremely complex concentrates encountered in flavour research laboratories, and for their identification it was considered necessary to use fast-scanning mass spectrometry coupled directly to high resolution gas chromatography. Under these conditions, the g.c. data obtained can compensate for any loss in accuracy due to deviations from the truly quantitative response of the mass spectrometer under fast scanning conditions.

An inexpensive fast-scanning mass spectrometer (Evans and Collins, see Paper 2.1), with a mass range of 25–300 u has been specially designed to operate with a g.c. inlet. This unit is compact and simple to operate. The chemist can, therefore, install it in his own laboratory where full use can be made of the existing undisturbed g.c. techniques, already developed for his special needs.

A brief description is given of the experimental conditions adopted and some preliminary results are reported.

2.2.2. EXPERIMENTAL

The ideal combination of these two techniques should allow the gas chromatograph to optimize the separation of components of the complex mixture and provide high efficiency and accurate retention data. The mass spectrometer would then scan rapidly over the mass range, so that the changing concentration of sample in the gas chromatographic peak does not affect the quantitative relationship between the various fragmentation ions present in the spectrum. Finally, the pump system should be capable of clearing the sample at a rate comparable to the period of peak emergence from the gas chromatograph, and leave no 'memory' in the mass spectrometer inlet and source. The following g.c.–m.s. experimental conditions were found to comply with these requirements.

2.2.3. GAS CHROMATOGRAPHY

For most complex flavour mixtures, temperature programming of the g.c. column is necessary to provide good separation and still retain high efficiency. The programme rate can be adjusted over a wide range on most commercial instruments and the correct selection of programme rate must be made, bearing in mind the requirements of the mass spectrometer. A fast programme rate produces highly efficient peaks but reduces separation. A slow programme rate provides good separation but often the peaks are broad and, therefore, at a low concentration per unit volume of carrier gas. A practical compromise gave moderate separation and an intermediate concentration. Providing there is some degree of separation between the components eluted from the g.c. column, then useful mass spectral data can be obtained by fast scanning at several points over the g.c. peak. The value of the mass spectral information will increase to a maximum with complete separation.

Support-coated open tubular (s.c.o.t.) capillary columns have been developed[3] recently and are now commercially available in this country. They have properties intermediate between those of a conventional packed column and the capillary column. They retain, to some extent, the high loading capacity of a packed column and the high resolution of a capillary column, and are therefore ideal for g.c.–m.s. The results quoted below were obtained using a s.c.o.t. column with diethylene glycol succinate (d.e.g.s.) as the stationary phase.

The carrier gas used was helium and a flow rate of 4 ml/min was found to be optimum for the 100 ft s.c.o.t. column. The emerging column gas flow was split in the ratio 1 to 10 between the g.c. flame ionization detector and the m.s. coupling tube. This split did not affect the quality of the g.c. record. The m.s. coupling tube was an 18 in. length of 0·02 in. internal diameter, stainless steel capillary, coated internally with silicone gum rubber (SE 30), to avoid absorption on the bare steel surface.

2.2.3.1. Sample Handling

A multi-purpose injection system was designed for the Pye Series 104 temperature programmed g.c. unit, already available in the laboratory, so that vapour, liquid or trapped samples[4, 5] could be injected. In the present study, 2 kg of potatoes were allowed to boil for 2 h under a water-cooled reflux condenser. The relatively water-free volatile substances which accumulated in the space above the cooled area of the condenser were drawn off into a concentration trap[4, 5]. After collection for 15 min, the trap was disconnected from the condenser and transferred to the carrier gas inlet line. Ohmic heating of the stainless steel trap was used to transfer the volatiles into the first few inches of the g.c. column, bent into a *U* shape and cooled in liquid oxygen. When the operator was ready to begin the analysis, the coolant was removed and the temperature-programming oven door closed. This technique minimizes peak spreading due to sample injection into relatively large volumes of carrier gas.

2.2.4. MASS SPECTROMETRY

The g.c.–m.s. coupling tube, carrying approximately 3·5 ml/min carrier gas flow was connected via a *T* piece to the m.s. inlet chamber and the odour outlet port (see below) so that a gas flow rate of 2·0 ml/min passed into the m.s. inlet chamber while the remaining flow was allowed to escape into the laboratory atmosphere.

The 'effective' volume of the m.s. inlet chamber was reduced by carefully positioning the inlet capillary tube close to the glass leak into the analyser. The residence time of the sample component in the inlet chamber was controlled by adjustment of a valve which partially closed the line to the inlet chamber pumping system. As soon as any peak, due to carbon dioxide or solvent present in large amounts (greater than 100 μg), had passed through the inlet chamber, the valve was closed sufficiently to produce a pressure of 10^{-5} torr in the analyser. The valve was then left at this position for the rest of the experiment.

The spectra were produced by scanning from 25 u to 300 u in 1 s, monitored visually with an oscilloscope, and permanent records were made at the appropriate times, with a u.v. oscillograph. The details of the mass spectrometer instrumentation are covered by Evans and Collins in Paper 2.1.

2.2.4.1. The Odour Port

The mass spectrometer could only handle a flow rate of helium gas of approximately 2 ml/min whereas the optimum flow rate for the g.c. column was 4 ml/min. The excess flow was utilized in two ways:

(*i*) To allow for changes in flow rate through the column during the programming.

(*ii*) To provide a flow of sample into the laboratory atmosphere, through the length of heated capillary tube, so that an experienced observer could smell the eluate and give odour descriptions (o.d.) whenever possible.

2.2.5. RESULTS

The combined information of g.c. retention data, mass spectrum and o.d., has been used to characterize the volatiles produced from cooking vegetables, raw and cooked chicken, and for model system studies in the flavour chemistry section in general.

A typical gas chromatogram obtained is shown in *Figure 2.2.1* and the combined data in *Table 2.2.1.* The peaks are numbered in order of elution and these numbers correspond to the peaks listed in *Table 2.2.1.* Consideration of *Figure 2.2.1* and *Table 2.2.1* together shows the combination of the three sources of information for the identification of the volatile substances contained in the mixture.

To illustrate more fully the approach to the problems of identification, the g.c. peaks were grouped under the following headings:

(*a*) Well-resolved g.c. peaks containing only a single compound based on the mass spectrum and odour description.

(*b*) Single or multiple g.c. peaks containing two or more components, slightly resolved so that two or more reliable, fast-scan mass spectra were available.

(*c*) Single g.c. peaks containing either a single unknown component with an unrecognizable mass spectrum or two more components completely superimposed, so that mass spectral data were extremely difficult to understand.

(*d*) G.C. peaks on which no m.s. were recorded.

(*a*) Peaks 1–3, 6–9, 13–15 and 24 were well-resolved g.c. peaks from which only single mass spectra were obtained. The majority of these compounds were readily identified by comparing their mass spectra with known data[6, 7] and were later confirmed by

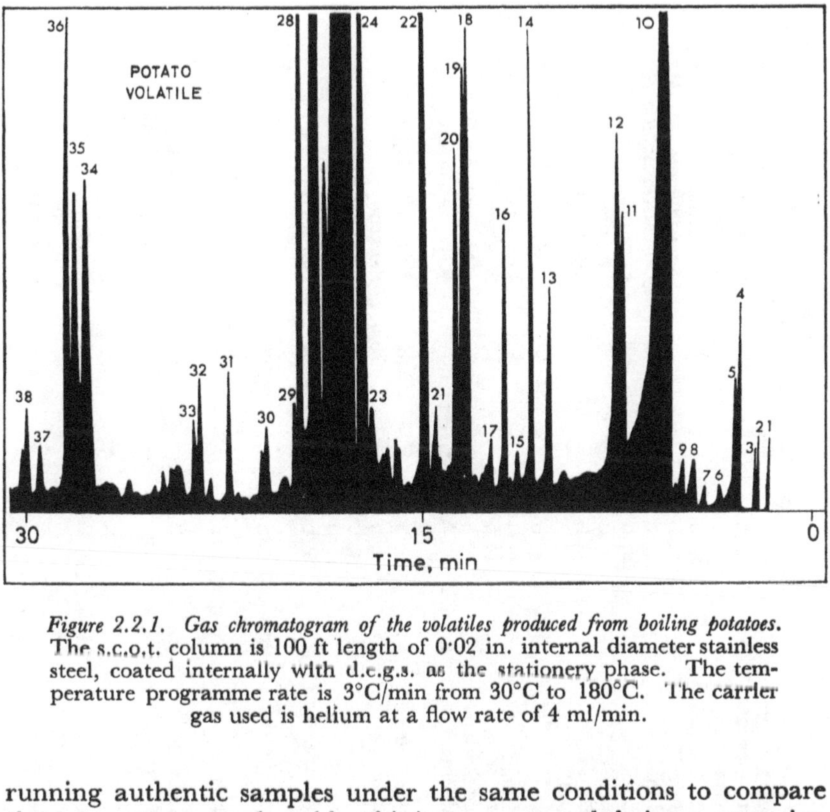

Figure 2.2.1. Gas chromatogram of the volatiles produced from boiling potatoes. The s.c.o.t. column is 100 ft length of 0·02 in. internal diameter stainless steel, coated internally with d.e.g.s. as the stationery phase. The temperature programme rate is 3°C/min from 30°C to 180°C. The carrier gas used is helium at a flow rate of 4 ml/min.

running authentic samples under the same conditions to compare the mass spectra produced by this instrument and their g.c. retention times. The odour descriptions given were characteristic of those of authentic samples, and even the 'no odour' comments were correct for peaks 3, 6 and 9 because these substances have very high olfactory threshold values and higher concentrations would be required for odour detection. Many of these substances have been reported previously in the odour of potatoes[8].

(*b*) Peaks 4 and 5, 10, 11 and 12, 18, 19 and 20 and the group 25–28 demonstrate the types of problem encountered with complex mixtures. Peaks 4 and 5, partially separated on the g.c. column, showed two distinct mass spectra of acetaldehyde and dimethyl sulphide. Peak 10 contained two components. Methanol was the

Table 2.2.1. The combined data obtained on the gas chromatographic peaks shown in Figure 2.2.1

G.C. peak number	*Retention time (min and s)*	*Odour description*	*Mass spectral data*	*Molecular weight*	*Identification*	*Comments*
1	1.40 [c]	Hydrogen sulphide [e]	a, b	34	Hydrogen sulphide	d
2	1.55 [c]	Sulphurous [e]	a, b	48	Methanethiol	d
3	1.59 [c]	No odour	a, b	74	Diethylether	d
4	2.38 [c]	Acetaldehyde [e]	a, b	44	Acetaldehyde	d
5	2.50 [c]	Sweet sulphide [e]	a, b	62	Dimethyl sulphide	d
6	3.40 [c]	No odour	a, b	68	Furan	
7	4.15 [c]	Aldehyde [e]	a, b	58	Propanal	d
8	4.35 [c]	Slightly brown, aldehyde [e]	a, b	72	2-Methylpropanal	d
9	5.00 [c]	No odour	a, b	58	Acetone	d
10	5.30 [c]	No odour	a, b	32	Methanol	d
10	5.30 [c]	No odour	a, b	70	Methacrolein	
11	7.12 [c]	Meaty, aldehyde [e]	a, b	86	2-Methylbutanal	d
11	7.12	Meaty, aldehyde	Possibly 2,4-hexadienal		R.T. not correct	More sample required
12	7.24 [c]	Meaty, aldehyde [e]	a, b	86	3-Methylbutanal	d
12	7.24 [c]	Meaty, aldehyde	a, b	78	Benzene	
13	9.55 [c]	Buttery [e]	a, b	86	2,3-Butanedione	
14	10.30 [c]	Low odour	a, b	136	α-Pinene	
15	11.10 [c]	Sweet, sulphide, light furan [e]	a, b	94	Dimethyl disulphide	a
16	11.35	Onion, garlic	206, 96, 133 (major ions)	206?	Unknown	
17	12.07	No odour	No spectrum		Unknown	More sample required
18	12.55 [c]	Burnt, buttery [e]	a, b	100	2,3-Pentanedione	
18	12.55	Burnt, buttery	Suggests terpene	136	Unknown terpene	change of phase required
19	13.03 [c]	Green [e]	a, b	100	Hexanal	d
20	13.27	No odour	91, 92, 119	134	Substituted benzene ($C_{10}H_{14}$)	R.T. suspicious
21	14.15	Metallic	b	106	Dimethylbenzene	R.T. suggests *o*-xylene
22	14.30	Low odour	a, Terpene	136	Unknown terpene	R.T. very similar to sabinene

23	16.35	Privet green	Confused		Unknown	More sample required
24	16.50 [c]	Potato water	a, b	136	Limonene	Odour, possibly mixture of peaks 23 and 24
25	17.20 [c]	Nasty green, laundry	b	88	3-Methylbutanol	
25	17.40	Nasty green [e]	b	88	1-Pentanol	Check R.T.
26	18.20	Onion	43, 58 only		Unknown	More sample required
27	18.43	Hay, cut grass	Suspected 2,4-nonadienal or pentyl cyclopentene base ion 81	138	Not conclusive	Change of phase required
28	19.10 [c]	No odour	a, b	120	Mesitylene	
28	19.10	No odour	Terpene?	136	Unknown	Further study of the monoterpenes required
29	19.24	Garlic, shoe polish	31, 51, 78		Unknown	
30	20.35 [c]	Methional [e]	No spectrum		R.T. agrees with methional	May require concentrate from 100 kg potatoes
31, 32, 33, 37, 38, 34 and 35	27.20, 27.42		Spectra suggest substituted benzenes			More sample required
36	28.10	Musty potato sacks	57, 29, 43, 41, 27, 55, 39		Unknown	Change of phase to check purity
39	No peak, 30.15	Damp musty potatoes	No spectrum		Unknown	More sample required

(a) Comparison spectrum obtained on authentic samples.

(b) Agrees with published data [6,7].

(c) Agrees with retention time of authentic sample.

(d) Already reported in potato flavour[8].

(e) Agrees with the odour description obtained on an authentic sample.

No odour indicates that the sample concentration was below the olfactory threshold concentration.

major component and a trace of methacrolein was readily distinguishable, superimposed on the methanol mass spectrum. Peaks 11 and 12, however, were seen to contain four components. Several scans over these two peaks enabled four mass spectra to be obtained. Three of these compounds were verified by subsequent g.c.–m.s. on authentic samples but one trace component, thought to be 2,4-hexadienal was disproved by the retention time. The three peaks 18, 19 and 20 contained three major components. Peaks 18 and 19 were identified by g.c. and m.s. and o.d. as 2,3-pentanedione and hexanal respectively.

Peak 20 was thought to be a substituted benzene, molecular weight 134. The spectrum of peak 18 also indicated the presence of a terpene but this has not yet been characterized.

Peak 25 was found to contain two isomeric C_5 alcohols and three fast-scan spectra over the peak helped to identify them. Peak 27 was thought to be 2,4-nonadienal by comparison with its mass spectrum and odour description but although no authentic sample was available to confirm this, the retention time for authentic 2,4-hexadienal indicated that this component was not likely to be the nine-carbon atom homologue. The base ion of this spectrum was 81 and the molecular weight was 138. A more likely structure would be that of pentyl cyclopentene but no retention data has yet been obtained. Peak 28 was tentatively identified as a trimethylbenzene, possible 1,3,5-trimethylbenzene (mesitylene), by retention time. There was also a trace of a monoterpene in this fraction.

(*c*) Peak 16 produced an unknown mass spectrum with a possible parent ion, as base peak, at 206 and other major ions at 96 and 133. Peak 36 gave a satisfactory mass spectrum but it could not be ascertained whether this was a single compound or not. For this particular peak the odour description makes it worthy of further detailed study. The procedure suggested for this would be to repeat the analysis on the d.e.g.s. column as before, except this time, a trap would be fitted to the odour port outlet while this component was being eluted; the trap would then be disconnected and its contents rechromatographed on a different stationary phase. If a second single g.c. peak produces a similar mass spectrum again, then this is most likely to be a single unknown component. If, on the other hand, two or more g.c. components are resolved, then their individual mass spectra may be more readily identified by comparison with known spectra.

(*d*) There were a number of g.c. peaks which contained insufficient material to enable mass spectra to be recorded; namely 17, 23, 26, and 39. For identification of these and the other trace components, a larger sample is required. This could be provided either by increasing the size of the total volatile sample extracted from the potatoes, or by using a helium separator to increase the concentration of these components in the helium carrier gas entering the mass spectrometer. The existing system, without a separator, requires approximately 0·1 μg of the aromatic hydrocarbons, sulphides, disulphides and diketones, which are chromatographed efficiently on this column, whereas up to a 3 μg sample of the alcohols and aldehydes was required before mass spectral data could be obtained. This comment does not take into account the differences in sensitivity of the mass spectrometer towards the different types of compound listed above.

An experiment has been planned to investigate the two factors contributing to this effect. In general, a different stationary phase could be selected to give better peaks and allow smaller quantities of these substances to be analysed. For the remaining peaks 21, 22, 29, 30–35, 37 and 38, some information was obtained on this phase but further and more detailed studies will be required before conclusive data can be obtained. Peak 30 was described as having the odour of methional, an important odour-contributing volatile, found in other foods. From earlier calculations it was realized that this substance would not be present in sufficient quantities from 2 kg of potatoes to produce a mass spectrum. The peaks 31–38 have only recently been concentrated sufficiently to produce any data. There appeared to be four substituted benzenes, a terpene and possibly an alcohol.

2.2.6. SUMMARY

The above results have been obtained from a series of experiments carried out to improve all aspects of the g.s.–m.s.–o.d. technique and not least the methods of sample concentration and elimination of water vapour. There still remains much work to be done to confirm the suggested identities and to characterize the unknowns in this fraction, and also to extend the study to the less volatile components of the flavour concentrate.

In general, the whole analysis must be repeated on at least one other stationary phase to confirm the identities suggested here and to produce more reliable information on the hitherto poorly separated regions of the chromatogram. This paper should, therefore, be regarded as an interim report, intended to demonstrate the methods employed rather than presenting any unequivocal identities.

My thanks are extended to my colleagues at the Food Research Institute; to Dr. D. G. Land for detailed discussion concerning the g.c. data used, to Miss N. M. Griffiths and Mr. J. E. Ayres for the odour descriptions made during these experiments, to Mr. T. C. Grey for confirmatory g.c. analysis and to Mr. J. Eagles who operated the gas chromatograph–mass spectrometer.

REFERENCES

[1] DREW, C. M., MCNESBY, J. R., SMITH, S. R. and GORDON, A. S. *Analyt. Chem.* 28 (1956) 979

[2] HOLMES, J. C. and MORRELL, F. A. *Appl. Spectrosc.* 11 (1957) 86

[3] PURCELL, J. E. and ETTRE, L. S. *J. Gas Chromat.* 4 (1966) 23

[4] SCOTT, R. P. W., FOWLIS, I. A., WELTI, D. and WILKINS, T., *Gas Chromatography*, A. B. LITTLEWOOD (Ed.), Institute of Petroleum, London, 1967

[5] SWAIN, T. and SELF, R. *Eur. Potato J.* 7 (1964) 228

[6] 'Index of Mass Spectral Data', *Am. Soc. Test. Mater. spec. tech. Publ. 365*, 1963

[7] CORNU, A. and MASSOT, R. (Eds.), *Compilation of Mass Spectral Data*, Heydon, London, 1966

[8] SELF, R., *The Chemistry and Physiology of Flavours*, H. W. SCHULTZ, E. A. DAY and L. M. LIBBEY (Eds.), AVI, Connecticut, 1967

2.3

A GAS CHROMATOGRAPH-MASS SPECTROMETER LINK-UP—RECENT DEVELOPMENTS

A. Vetsch

EVERY chemist, when confronted with a mixture of unknown composition, faces two main analytical problems:

(*i*) Separation of the different components.

(*ii*) Identification of these components.

These two problems were, for a long time, very difficult to deal with, although for generations chemists searched for and developed more and more sophisticated techniques for solving them. Finally, the application of physical methods to analytical problems lead to a first breakthrough in analytical chemistry. Today, methods like u.v., i.r. and n.m.r. spectroscopy, polarimetry, optical rotatory dispersion, gas–liquid and thin-layer chromatography, X-ray diffraction and mass spectrometry, in conjunction with modern advanced data-processing equipment, enable the chemist, not only to solve very difficult analytical problems, but to solve them within hours or days. Previously, the solution of such problems would have required weeks, or even years, of work. For a long time these physical methods were not regarded as classical chemistry but today they are a very important part of every chemist's education, especially that of the organic chemist.

Although gas chromatography, for example, is an excellent method for separating complex mixtures, it lacks considerably the ability to identify the separated compounds. On the other hand, very well-established identification techniques like u.v., i.r., and n.m.r. spectroscopy are generally not fast enough to follow the rapid separation provided by a gas chromatograph. However, within the last few years, the direct combination of gas chromatography (separation) and mass spectrometry (identification) has lead to what is probably the most powerful analytical technique today.

Perkin–Elmer, as one of the leading manufacturers of analytical equipment, became increasingly aware of the importance of this

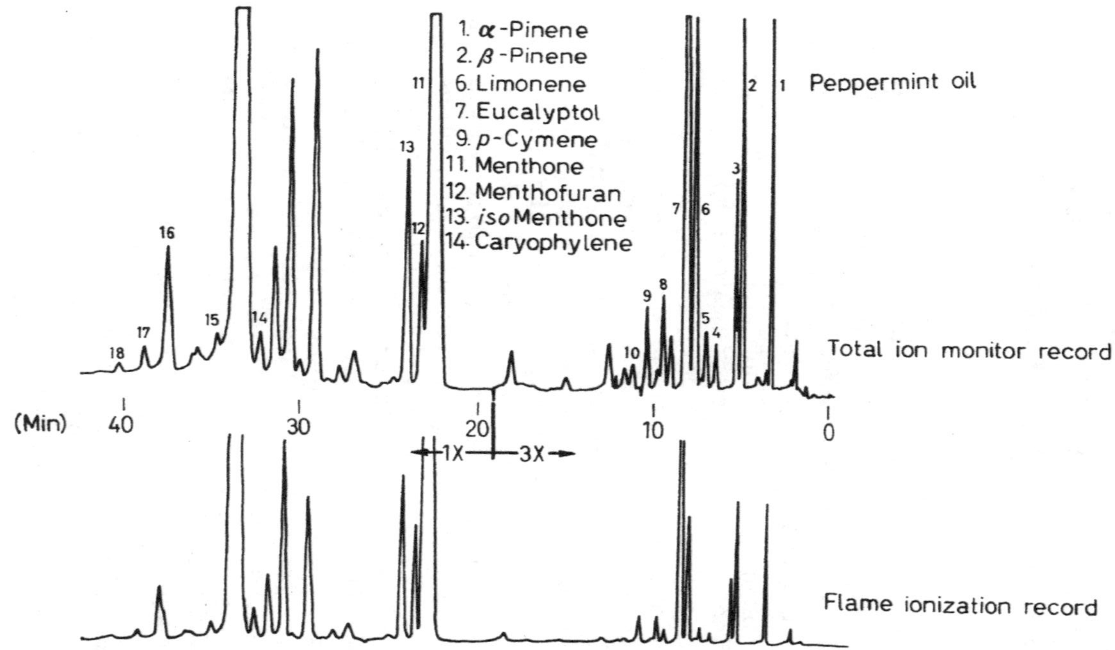

Column 0·02 in. i.d. support coated open tubular, Carbowax 1540. He carrier gas (8 lb/in²). Temp. 70°–150°C at 2°C/min. Sample 0·1 μl

Figure 2.3.1. Direct comparison of a chromatogram of peppermint oil obtained simultaneously by the flame ionization detector and the total ion monitor

technique, which has been pioneered, among others, by Beynon, Stenhagen, Ryhage, Holmes, Morell, Watson and Biemann. About 3 years ago, we started to connect our gas chromatographs to the Hitachi Perkin–Elmer RMU6 mass spectrometer. During the 14th Dallas meeting, Struck and Averill[1] reported the following results.

A Perkin–Elmer Model 801 gas chromatograph was coupled by a heated line and a Watson–Biemann separator, directly to the mass spectrometer. The RMU6 is a magnetic scanning sector-type instrument with a mass range of 2400, a resolution of 2500 and a scan speed of up to 1 s per decade. A split at the end of the column allowed the eluents to be fed to the flame ionization detector and to the mass spectrometer simultaneously, so that a direct comparison between the separation obtained at the end of the column and the separation found in the ion source could be made.

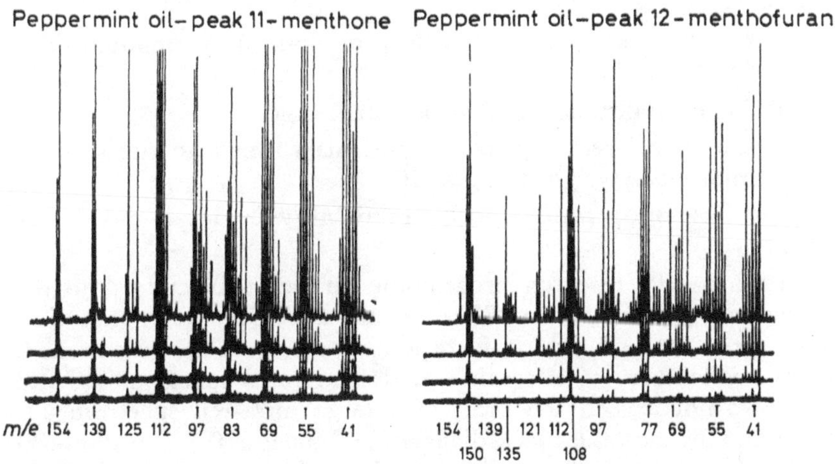

Figure 2.3.2. Mass spectra of incompletely separated menthone (peak 11) and menthofuran (peak 12)

Figure 2.3.1 shows a direct comparison of a chromatogram of peppermint oil obtained simultaneously by the flame ionization detector and the total ion monitor. The described system does not have any memory or background problems, as illustrated in *Figure 2.3.2.* Although peaks 11 and 12 were not completely separated gas-chromatographically, the large molecular peak *m/e* 154, of g.c. peak 11, is very small in the spectrum of g.c. peak 12. Despite the fact that the results were quite good with this combination (other results[5] are shown later) we felt that the performance of the Hitachi Perkin–Elmer mass spectrometer was unnecessarily high for this particular application, since generally about 80–90% of all samples in an organic chemistry laboratory do not have a molecular weight higher than 800. Also, for g.c. work, molecular weights above 500 are rather rare. Therefore, a mass spectrometer with reduced specifications (compared to the RMU6) would fulfil the needs of most organic chemists.

The most important capabilities of such an instrument may be briefly summarized as follows. It would provide:

(*i*) A sampling system for gases, liquids and solids.

(*ii*) Direct introduction of liquids and solids into the ion source.

(*iii*) Direct connection to a gas chromatograph.

(*iv*) Fast scanning ability, to cope with the separation obtained from packed, open tubular and support-coated, open tubular columns.

(*v*) Fast recording system with provisions for computer-aided data reduction.

(*vi*) Detectability in the nanogramme range.

(*vii*) Adequate resolving power and mass range, to run samples with a molecular weight of up to 1000.

(*viii*) Ease of operation and high reliability for day-to-day operation.

The intensive research programme on this subject lead to the development of the Perkin–Elmer Model 270, G.C.–D.F. analytical mass spectrometer[2], which incorporates all the above-mentioned, necessary features of a routine analytical mass spectrometer and still does not exceed the price of a large infra-red spectrometer. *Figure 2.3.3* shows the block diagram of Model 270. The standard equipment consists of a modified, high performance, Perkin–Elmer Model 900 gas chromatograph oven[3], which is heatable up to 350°C and accepts all types of stainless steel or glass column. The connection to the ion source is made by a Watson–Biemann separator[4], which can be heated separately, up to 400°C. The direct-introduction system allows the sample to be introduced directly through a vacuum lock into the ion source. The sample is held in a small quartz capillary which can be heated to 350°C in 10 min. The temperature is measured by a thermocouple placed adjacent to the capillary.

At 75°C the $(M-18)^+$ peak of cholesterol is less than 30% of the parent ion peak. A heated metal inlet system with a reservoir is

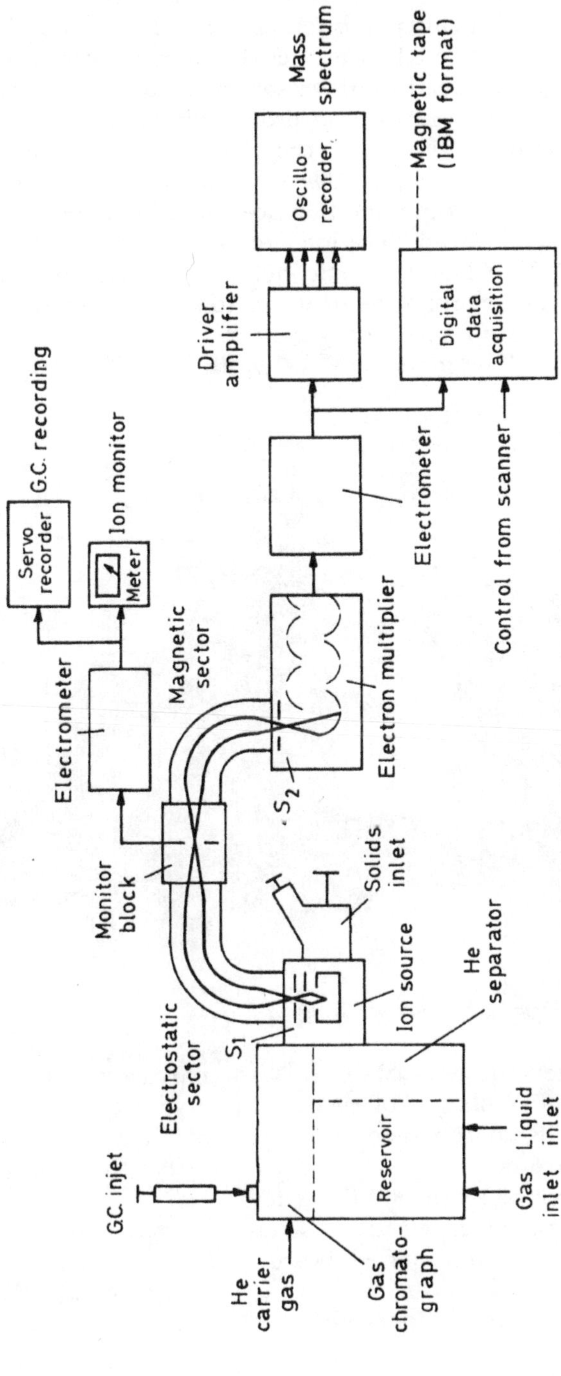

Figure 2.3.3. Block diagram of Perkin–Elmer Model 270 GC–DF *analytical mass spectrometer*

optionally available. The high emission current, electron bombardment, ion source is connected to a Nier–Johnson double focusing analyser with a 90°, 5 in. electrostatic sector and a 60°, 4 in. magnetic sector, corrected for first and second order abberations. The ion source uses magnetic collimation of the electron beam which has an energy adjustable between 5–100 eV. The source is designed in such a way that heavy contamination does not affect the performance or operation. The acceleration voltage can be set to 750, 1000, 1500, 2000 V. The magnetic field, which can be varied between 0 and 16 kG, enables the operator to scan a mass range of *m/e*, 2–1000 with a scan speed of one decade per second to one decade per 10 min. The ion source slit width is 0·002 in. whereas

Figure 2.3.4. Analyser section of Model 270

a variable collector slit is available. With narrow collector slit width (0·001 in.) a resolution of $m/\Delta m = 850$ with a 10% valley can be obtained. Resolution and sensitivity are of course interrelated in mass spectrometry so that one can always sacrifice some resolution for increased sensitivity if needed.

An electrode between the electrostatic and the magnetic analyser intercepts part of the undispersed beam and acts as a total ion monitor for adjusting the instrument, or as a g.c. detector. *Figure 2.3.4* shows the analyser of Model 270.

The ion detector is a 16-stage Ag–Mg electron multiplier. The output of the multiplier is fed into a fast amplifier. At sensitivity 1, with a $10^7\ \Omega$ input resistor, the bandwidth is 10,000 Hz with an input current of approximately 5×10^{-8} A. The 0·5 V output from the signal amplifier will produce full-scale deflection of the most sensitive galvanometer; for instance, when the Honeywell 2000 Hz (M 3300) galvanometer is driven by the built-in driver amplifier. Slower galvanometers can be used in order to gain higher sensitivity. The output of the driver amplifier can also be fed directly into the digital data acquisition system (available on option) which records the spectrum in digital IBM–formated form. A description of this system is given later.

Two complete vacuum systems pump differentially, the ion source and the analyser. The systems consist of a rotary pump and a water-cooled diffusion pump. Water-cooled baffles prevent back-

Figure 2.3.5. Model 270 GC–DF, *analytical mass spectrometer*

diffusion of oil vapours into the source and analyser. The entire system, which is protected against failure of water, vacuum, or line voltage, can be baked up to 300°C. The pressure in the system is measured by a cold cathode gauge. The vacuum in the clean system, without sample or carrier gas introduced, is normally below 10^{-7} torr. *Figure 2.3.5* gives an impression of the complete instrument.

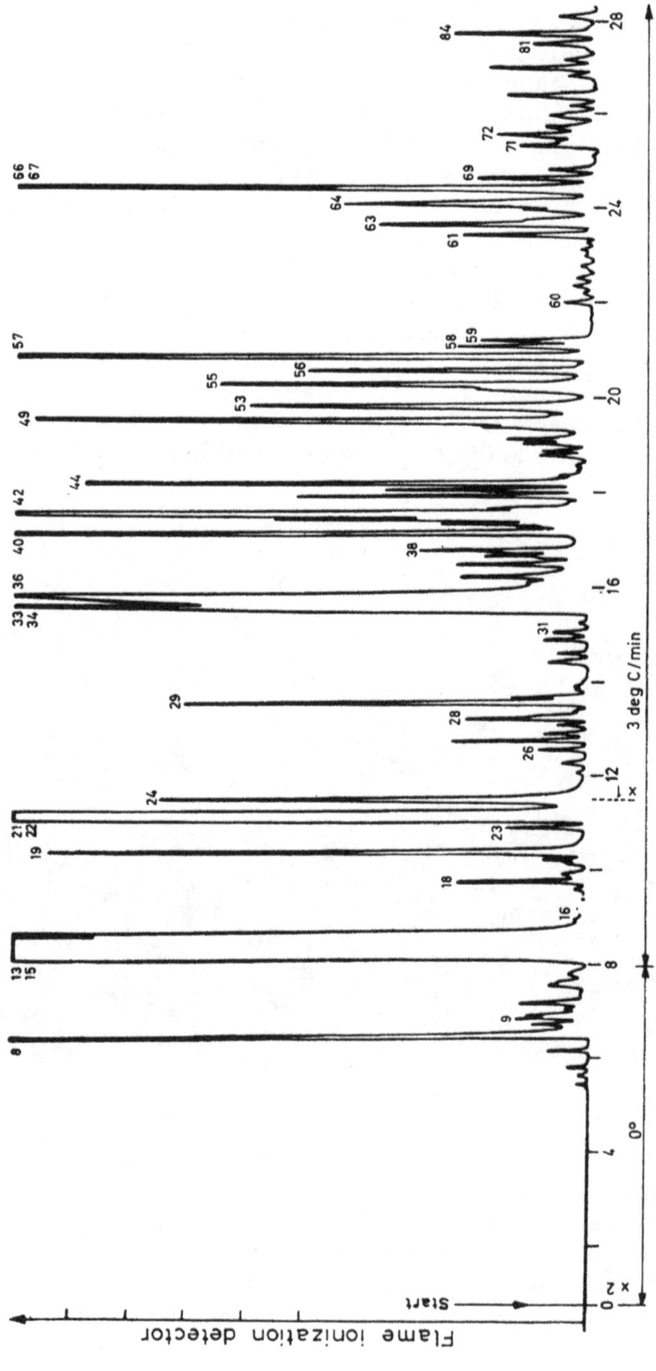

Figure 2.3.6. Gas chromatographic separation of constituents of cigarette smoke: (a) flame ionization detector. (After J. Völlmin, et al.[5]*)*

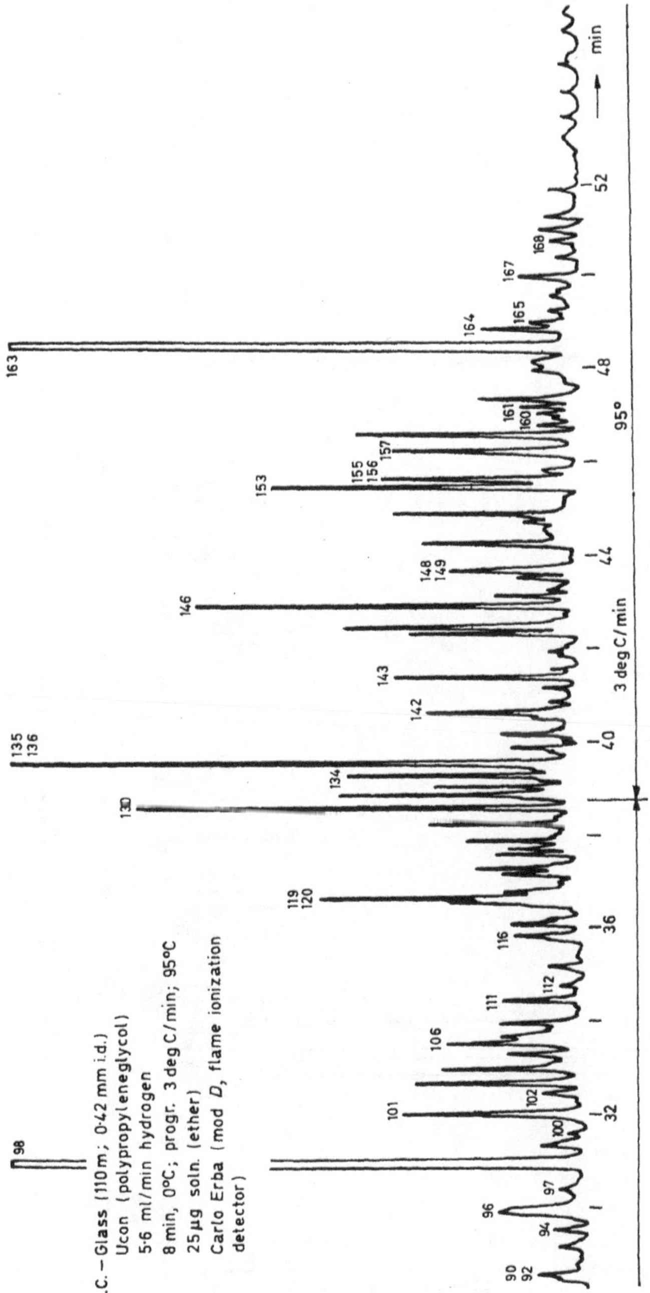

Figure 2.3.6. (a) continued

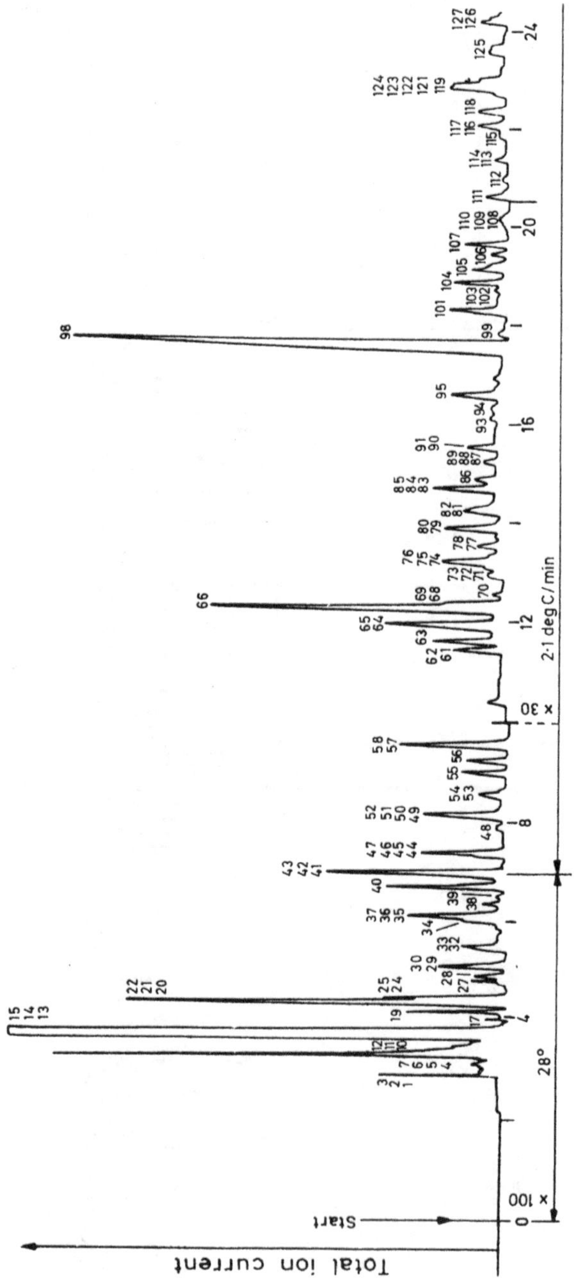

Figure 2.3.6. Gas chromatographic separation of constituents of cigarette smoke: (b) total ion current detector. (*After* J. Völlmin *et al.*)

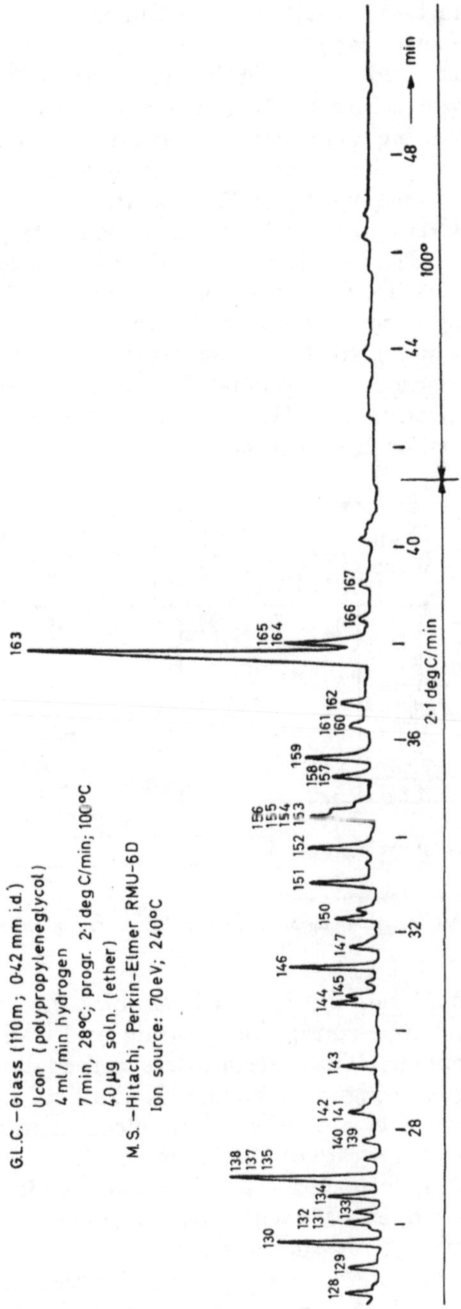

Figure 2.3.6. (b) continued

When a g.c.–m.s. combination is in full use, day after day, it can produce an enormous amount of spectra. A very good example was recently published by Völlmin, Omura, Seibl, Grob and Simon[5]. Grob[6], who is investigating cigarette smoke, could separate about 250 peaks using glass capillary columns. The chromatograms obtained with the flame ionization detector and the total ion monitor are shown in *Figure 2.3.6.* About 300 spectra were taken with the RMU6 during 1 day and then interpreted within about 3 weeks. One can easily imagine the amount of information that can be obtained from such an instrument and that very soon the data processing becomes a real problem.

Professor Biemann[7] from the Massachusetts Institute of Technology, who is a pioneer in computer-aided data processing of high-resolution, mass spectrometry data, evidently had similar problems. At the recent Pittsburgh conference, he presented a digital data

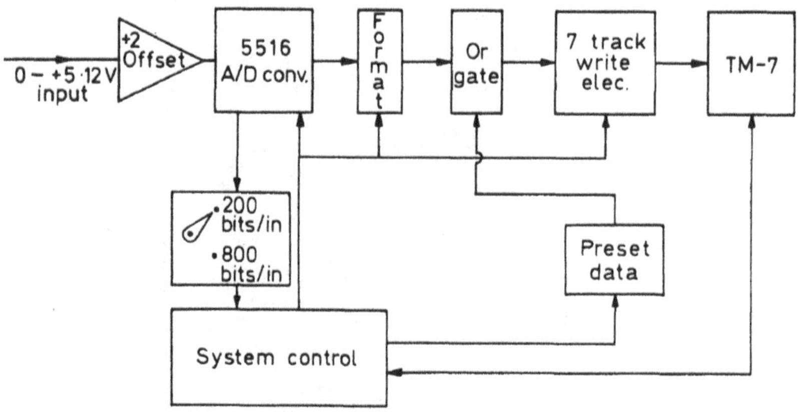

Figure 2.3.7. Block diagram digital data acquisition system after Biemann

acquisition system (*Figure 2.3.7*) which enables the mass spectroscopist to leave the time-consuming plotting and standardizing of spectra to the computer. This system works in the following way.

A 12 bit, analogue-to-digital converter (Radiation Inc.) is coupled to the output of the d.c. amplifier of the electron multiplier, and samples the signal with a rate of 3000 data points per second. This digital signal is then formated into IBM octal code and written onto the tape of an Ampex TM2 with a packing density of 200 bits/in. Additional circuitry allows such present data as operator, scan number and a 3/4 in. interrecord gap, to be written onto the tape. Furthermore, it controls the whole system by using either 'local' or 'remote' control. When the 'local' control is chosen, the digitizer and the tape recorder are actuated by start and stop switches, located on the panel of the instrument. When switched to 'remote' control, the entire system is controlled by the scan button of the mass spectrometer.

During every scan, about 6–9000 intensity values are digitized. This is equal to about 5–10 points per peak. This data is then processed by a computer which assigns the exact integral mass number and the intensity to every peak. In order to calibrate the system, a cycling scanner is used since it was found that after about five scans, the recording conditions become very reproducible (less than $\pm$ 0·2 u at mass 200). As a mass scale, a saturated hydrocarbon is used as an external standard. Furthermore, by introducing air, peaks 28 and 32 could easily be identified as a starting point. The computer is programmed in such a way that it first recognizes peaks 28 and 32 and then assigns the exact integral mass number to every peak. The intensities of peaks 28 and 32 are then set to zero and the intensities of all the other peaks are normalized in reference to the remaining most intensive peak (base peak). The computer then either prints out the spectrum, in tabulated form or as a standardized line spectrum. It can also be stored in the memory of the computer for further treatment.

When using an IBM 7094, processing a spectrum takes about 20 s. Although such a digital data acquisition system is still relatively expensive, when combined with a low cost mass spectrometer as described before, the total costs are not higher than the price one had previously to pay for a mass spectrometer alone. If one considers the time which can be saved by enabling the chemist to use a g.c.–m.s. combination to its full capacity, then such a system cannot be considered expensive.

I should like to thank Professor Biemann and Dr. Hites from the Massachusetts Institute of Technology for the information on the digital data acquisition system and Mr. E. Delany from our application laboratory for the spectra taken with Model 270.

REFERENCES

1 STRUCK, A. H. and AVERILL, W. *Am. Soc. Test. Mater. 14th A. Conf. Mass Spectrom. Allied Topics*, Dallas (1966)

2 NIER, A. O. and DELANY, E.B. *Am. Soc. Test. Mater. 15th A. Conf. Mass Spectrom. Allied Topics*, Denver (1967)

3 REDMOND, M. W. and CONDON, R. D. *Pittsburgh Conference on Analytical Chemistry and Applied Spectroscopy*, March, 1967

4 WATSON, J. T. and BIEMANN, K. *Analyt. Chem.* 37 (1965) 844

[5] VÖLLMIN, J., OMURA, I., SEIBL, J., GROB, K. and SIMON, W. *Helv. chim Acta.* 49 (1966) 1768
[6] GROB, K. *J. Gas Chromat.* 3 (1965) 52 and *Beitr. Tabakforschung* 3 (1965) 243
[7] HITES, R. A. and BIEMANN, K. *Pittsburgh Conference on Analytical Chemistry and Applied Spectroscopy*, March, 1967

2.4

A SMALL 180° DEFLECTION PARTIAL PRESSURE-TOTAL PRESSURE GAUGE FOR VACUUM SYSTEM DIAGNOSIS

M. Day and S. Evans

2.4.1. INTRODUCTION

THE comparatively recent use of cheap, low resolution mass spectrometers for residual gas analysis and process monitoring in vacuum systems, has improved many vacuum techniques and processes. A mass spectrometer enables the composition of the residual gas to be determined and this information permits a more realistic interpretation of the total pressure, compared with the nitrogen equivalent given by a total pressure gauge. The source of the residual gas can often be identified from an analysis of its components, for example it is possible to distinguish between a real and a virtual leak. A mass spectrometer can also tell whether a particular run has been repeated under nearly identical conditions, as well as monitoring the partial pressure of any particular gas.

Until comparatively recently, few vacuum workers had applied mass spectrometers to residual gas analysis because, in the past, most of the instruments commercially available were designed without any special emphasis on vacuum physics requirements. The most important limitations of these instruments have been expense, slow scan times and sensitivity to contaminating films causing unreliability in dirty conditions. These early mass spectrometers contributed significant gas loads that could both limit the ultimate pressure obtainable in the vacuum system and give a high background reading. This gas contribution can be troublesome and explains some of the success of the Omegatron which, considering its price, has rather inadequate resolution and sensitivity but minimum outgassing, as the internal electrode assembly of the head can be eddy-current heated to 800°C.

For a mass spectrometer to be suitable as a residual gas analyser it should have the following specification:

(*i*) The resolution should be sufficient to enable the more common residual gases to be analysed and, in any case, should be such that mass 44, i.e. carbon dioxide, can be distinguished from mass 43 which is a characteristic peak for many organic compounds.

(*ii*) The partial pressure sensitivity should be as high as possible, giving a lower limit of detectability of the order of 10^{-10} torr or less.

(*iii*) The time required to scan the mass range of the instrument should not be too long, as the composition of the residual gas may vary quite rapidly, particularly if any sorption or desorption processes are taking place within the vacuum chamber.

(*iv*) The internal construction of the head must be robust and reassembly after cleaning or replacement of filament, etc., must be a reasonably simple operation.

(*v*) The head must contribute a minimum gas load. If the head does outgas significantly, the mass spectrometer itself may limit the ultimate pressure obtainable in the system. To ensure that the gas being analysed is representative of the residual gas, there must be a minimum of outgassing and the pumping speed of the head must be small. It follows that the head should be bakable and as small as possible, as well as having limited active surface areas and the materials used should have low outgassing properties as well as being chemically resistant to corrosion, etc.

(*vi*) The head should not be too sensitive to variations in any of the operating parameters. In particular, for magnetic deflection instruments the performance should be reasonably insensitive to the magnet position, as the magnet will usually have to be removed during any baking procedure, and realignment of the magnet on certain instruments such as the Omegatron, can be time consuming. If the instrument's performance is susceptible to small variations in operating potentials, these potentials would have to be highly stabilized or closely monitored.

These principles were applied to the design of a mass spectrometer for use as a partial pressure gauge, with application to both residual gas analysis and vacuum process monitoring.

2.4.2. GENERAL DESIGN CONSIDERATIONS

To obtain a cheap, robust and small instrument a 180° magnetic deflection design was used and constructed using vacuum-engineering principles. The head was made as small as possible to minimize outgassing and magnet size, whilst being sufficiently rigid to be unaffected by such vibrations as a rotary pump may generate. The associated short path of the ions enables the gauge to be operated at relatively high pressures before scattering adversely affects the linearity of the response. The ion path radius of 1·5 cm was selected to give a resolution of over 40, using slit sizes that could be easily manufactured and give acceptable sensitivity. The head was bakable to 400°C and the material used was principally stainless steel which, although not ideal, is a reasonable compromise in price. The electronics had to be cheap and, for this reason, a standard ionization gauge amplifier was used. The slowest response time of this amplifier was 0·5 s, using a 10^{12} Ω resistor, which is compatible with the fastest pen-and-ink potentiometric recorders enabling a scanning speed of about 1 u/5 s to be used before any clipping of the peaks occurred. It was not anticipated that the instrument would be used for fast scanning so the emission could be controlled by temperature rather than space-charge[1]. To reduce insulation problems it was decided that the maximum voltage used would be 500 V, which would correspond to mass 2 with a magnetic field strength of approximately 3 kG.

2.4.3. THE CONSTRUCTION OF THE HEAD

The construction of the internal electrode structure is shown in *Figure 2.4.1.* The method of constructing the head was similar to that used in a 60° mass spectrometer described by S. Evans (see p. 78). It consisted of jig-drilling all the components and mounting them on precision ceramic or stainless steel tubes, using either ceramic or stainless steel washers as spacers. The tubes were held in position with 10 BA studding and the ion source assembly was kept under tension using nimonic springs which allowed for differences in thermal expansion during baking. The ion beam defining slit and the collector slit were spark eroded in 0·010 in. thick stainless steel and mounted on the precision tubes. This method of assembly facilitated stripping down the head for cleaning, filament replacement or changing slit sizes, as no jigs were required for reassembly.

The ion beam defining slit could be connected to the amplifier and thus the ion current leaving the ion source, but not entering the analysing chamber, could be measured. One of the parameters to which this current is proportional is the total pressure, which could thus be measured as a nitrogen equivalent and, as the ion beam defining slit could not see the filament, there was no X-ray limit. The partial pressure collector consisted of a flat metal plate mounted on a rod, as any secondary electrons produced could not escape because of the high magnetic field, making a Faraday collector unnecessary. A filament shield was connected to the ion box and a repeller plate was mounted on the side of the box opposite to the ion exit slit.

The analysing chamber was an open structure, to reduce the surface areas in the head, although by doing this the use of the same internal electrode structure in a glass envelope was precluded. The

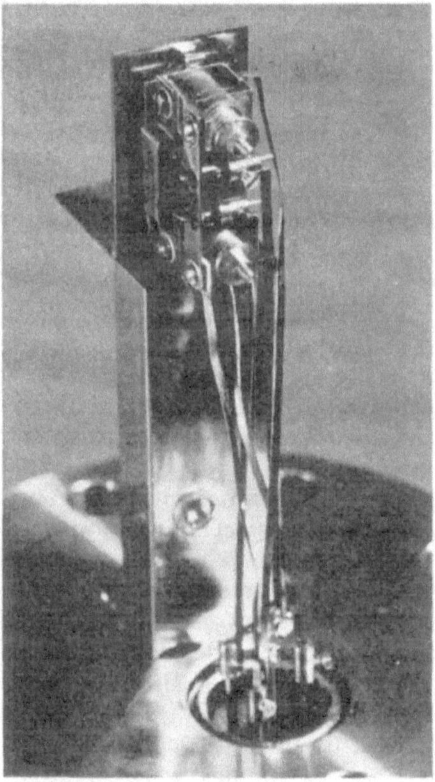

Figure 2.4.1. Construction of the internal electrode structure

open structure increased the distance from the ion beam to the nearest metal parts on which contaminating films may be formed and subsequently become polarized. The further that the ion beam is from these films the less is any affect on the performance of the instrument.

2.4.4. THE ELECTRONICS

The electronics used with the head consisted of:

(*i*) An h.t. supply and emission controller.

(*ii*) The amplifier.

These are shown in *Figure 2.4.2* which shows both the electronic controls and the assembled head.

The h.t. supply used a motorized helipot to give an inverse law output and hence a linear displacement of mass when using a potentiometric recorder. The voltage range of the output was from about 600 V to less than 10 V giving a mass range of 2–105 u. It was found that the resolution at mass numbers 2 and 4 was high enough to make a potentiometric recorder, with a response time of 0·5 s, clip the peaks at a scanning speed that gave a reasonable total

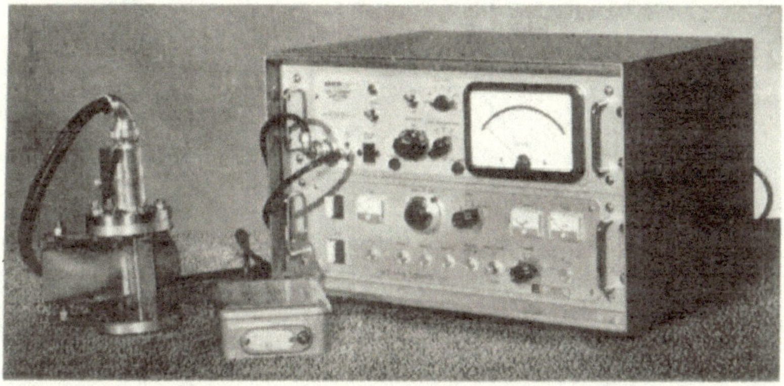

Figure 2.4.2. Complete instrument including electronics

scan time. This difficulty was overcome by putting in switch positions for masses 2, 3 and 4. It was visualized that one set of electronics may be required to operate several heads and to enable the duodial to read mass numbers directly. For different head constants a control was incorporated into the unit that could vary the output voltage by a small factor.

The emission was stabilized by a filament temperature controller which had switch positions giving four different electron beam currents of nominally 10, 30, 60 μA and 100 μA. The filament current, the box current and the electron beam current were monitored by meters. The electron accelerating potential was fixed at +76 V. The maximum filament current could be set by means of a preset potentiometer and a non-surge device prevented current surges when first switching on the filament. The repeller voltage was controlled by a potentiometer from +4 V to −3 V with respect to the ion box. The amplifier was a standard Edwards Model 2 (described by Steckelmacher[2]), with a noise level of 4×10^{-15} A, and the minimum current that could be easily detected on a potentiometric recorder during a scan, as opposed to sitting on a peak, was 5×10^{-15} A.

2.4.5. PERFORMANCE OF THE INSTRUMENT

The minimum partial pressure that the instrument could detect was 2×10^{-10} torr, nitrogen equivalent, and the minimum total pressure was 2×10^{-11} torr, nitrogen equivalent, both of which corresponded to an ion current of 5×10^{-15} A. The resolution gave

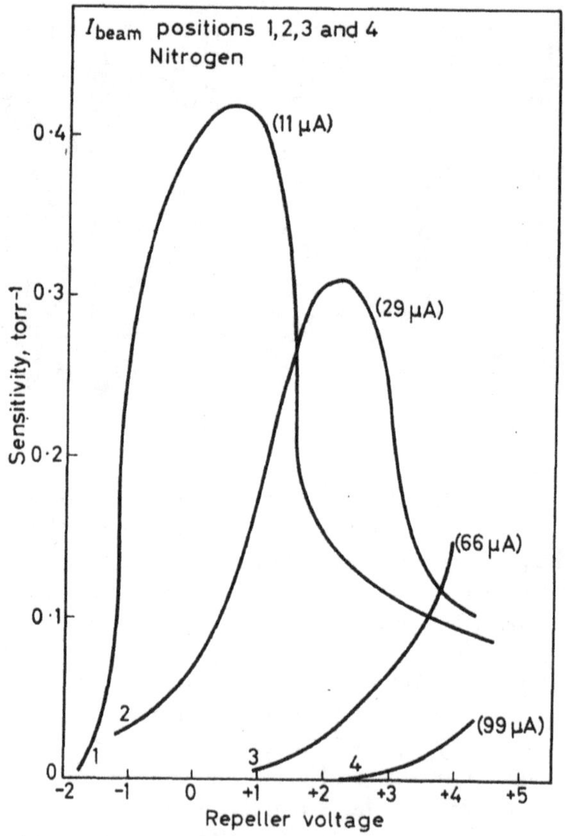

Figure 2.4.3. Variation of sensitivity with repeller voltage for various emission levels

better than a 10% valley at mass 44 and a useful spectrum was obtained for the whole mass range. The mass range of 2–105 u enabled mercury to be detected as doubly charged ions. The scan speed was about 10 s/u although this could be reduced, if a potentiometric recorder with a response time of $\frac{1}{2}$ s was available, to 5 s/u. The latter would give a scan time of about 8 min for the whole mass range, or 3 min to scan masses 2 and 12–45 which is often the only section of the scan of interest. The scanning time could be reduced even further but this would result in peak clipping and loss of resolution.

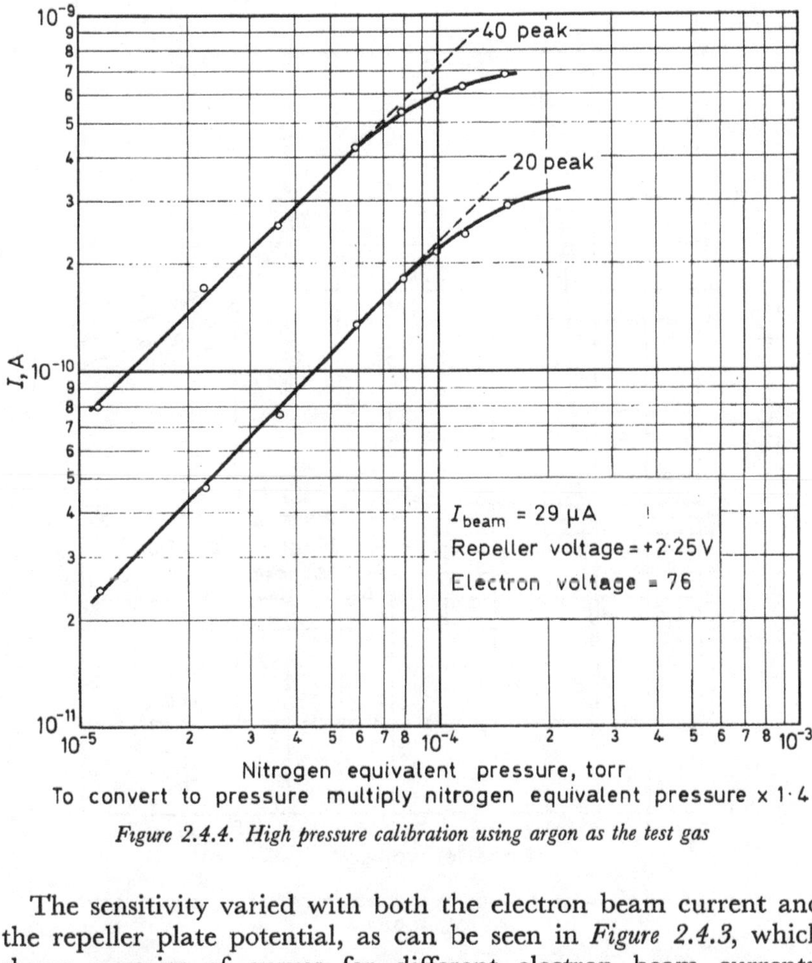

Figure 2.4.4. High pressure calibration using argon as the test gas

The sensitivity varied with both the electron beam current and the repeller plate potential, as can be seen in *Figure 2.4.3*, which shows a series of curves for different electron beam currents, corresponding to the four switch positions on the emission controller. The sensitivity used here is defined by the equation, $S = i^+/p \times i^-$. The variation with the repeller plate was not rapid enough to cause any great difficulty, although the value of the repeller plate setting that corresponded to maximum sensitivity for a given emission, was somewhat mass dependent. The maximum sensitivity occurred

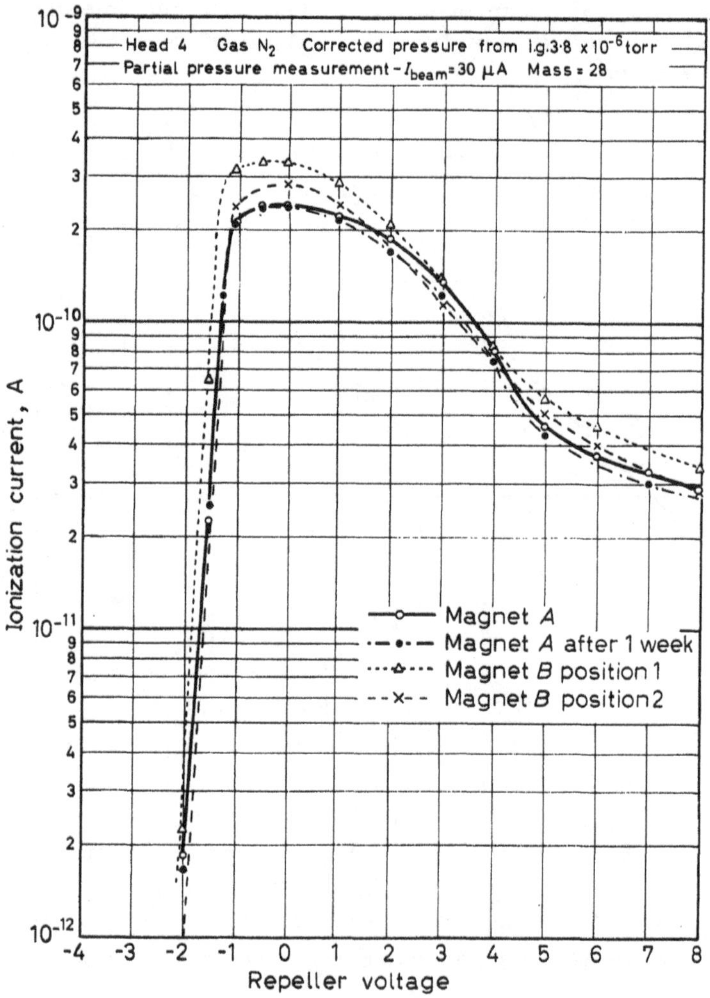

Figure 2.4.5. Variation of sensitivity with repeller voltage for various magnets and magnet positions

for an electron beam current of 10 μA, which corresponded to the lowest filament temperature. However, the 30 μA setting gave a lower partial pressure detectability than the 10 μA setting because, although the sensitivity was reduced by 25%, the emission was increased by about 30%. The sensitivity and resolution tended to fall with increasing emission except for light gases, especially hydrogen, for which the sensitivity remained relatively constant. Thus, using an I_{beam} of 100 μA, enabled the instrument to detect very small partial pressures of hydrogen and helium, making the gauge suitable for use as a very sensitive leak detector. The total pressure sensitivity increased with the emission but also varied with the applied h.t. and a calibration curve was necessary, particularly if both total and partial pressures were to be measured simultaneously, using two amplifiers.

The filament material was rhenium, which gave rise to minimum chemical fragmentation of the sample, low gas contribution and an acceptable evaporation rate.

The performance of the gauge at high pressures of argon is shown in *Figure 2.4.4*. The high pressure limit of the gauge was approximately 10^{-4} torr but the response became non-linear at pressures of about 6×10^{-5} torr and above, depending on the gas being examined.

The reproducibility of the instrument was good and the variation of the performance, with two magnets in different positions, is shown in *Figure 2.4.5*. The difference in sensitivity varied by up to 30% for the two magnets, but for different magnet positions with the same magnet, the maximum variation was 10% and was usually considerably less.

2.4.6. ANALYSIS OF SPECTRA OBTAINED FROM THE 180° PARTIAL PRESSURE GAUGE

The analysis of the spectra obtained with the residual gas analyser was a comparatively easy task, usually simplified even further for most vacuum workers, by the fact that they want qualitative rather than quantitative information. The cracking patterns and relative sensitivities of all the common gases have been experimentally determined for the 180° partial pressure gauge but the gas inlet system available was not suitable for calibrating with water vapour, with which it is particularly difficult to obtain accurate results. The results obtained compared favourably with similar types of instrument such as the AEI MS10.

With any instrument, the cracking pattern will vary with the filament temperature and if temperature control emission is used, this may lead to problems if the filament becomes poisoned to any extent.

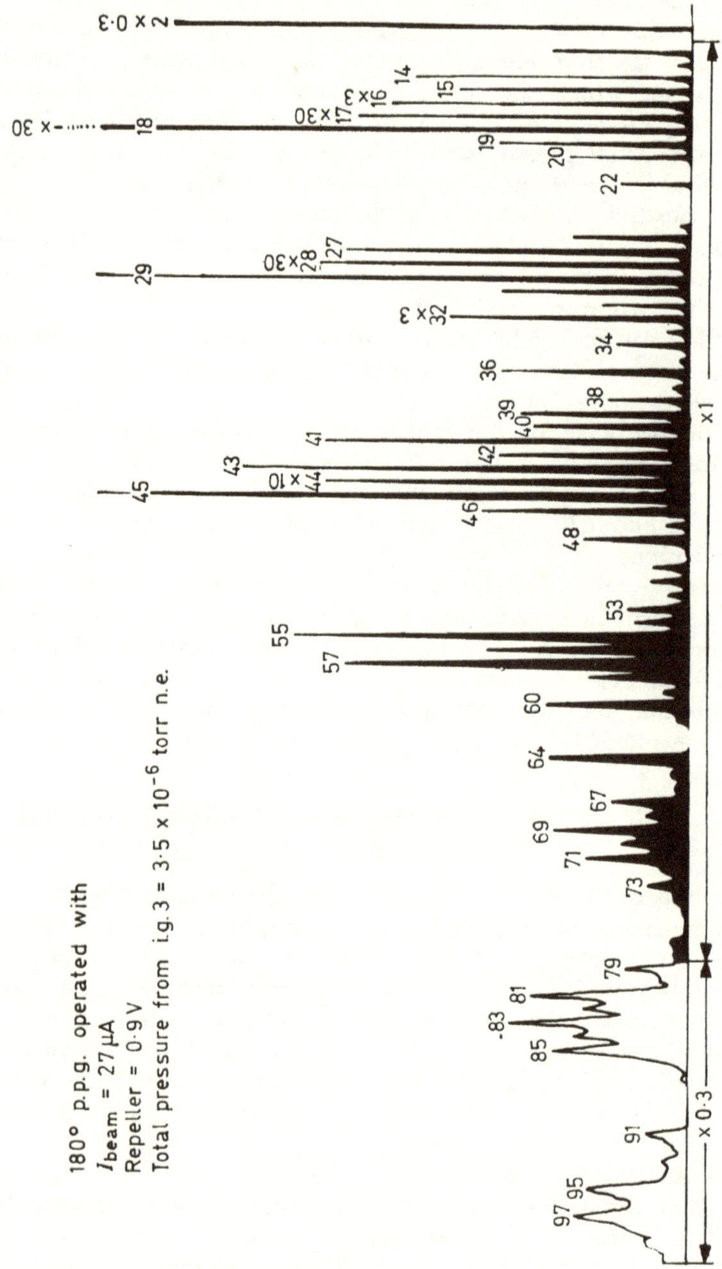

Figure 2.4.6. Mass spectrum of an unbaked system contaminated with rotary pump oil

The spectra obtained often contain many components, but these usually have simple and easily identifiable cracking patterns and the procedure for analysis of such spectra is well documented (Craig and Harden[3]). There is an ambiguity at peak 28, which may arise from nitrogen, carbon monoxide or ethylene. This can be resolved by examination of the fragment peaks.

2.4.7. APPLICATIONS AND EXAMPLES OF USAGE

This partial pressure gauge has proved suitable for general vacuum use and can often replace an ionization gauge to considerable advantage. The partial pressure gauge gives a more accurate pressure indication and can identify gas sources indicating whether there is a real or virtual leak. The head can also be used as a very sensitive leak detector with a variety of search gases, especially hydrogen and helium. If one of these instruments is fitted to a vacuum plant, it will give information on the gas load which can be used to reduce cycling times, telling, for example, whether a Meissner coil would be useful. The gauge can be used to monitor any process precisely and a 180° partial pressure gauge is being used on a prototype coating plant, to monitor the oxygen and water vapour content of the residual atmosphere, when semi-conductor oxide resistance films are being deposited. This has lead to a greater understanding of the deposition mechanism and a more reliable product.

The sample spectrum in *Figure 2.4.6* demonstrates the performance of the head and shows that the 10% valley occurs above mass 44. The heading of this figure is misleading. The head was on a system where the rotary pump pressure rose above the critical backing pressure of the diffusion pump. When the system was pumped-down again and the head switched on, this spectrum was obtained. However, the head had just been cleaned and had not been baked, so undoubtedly a lot of the peaks are due to cleaning fluids, etc., which could only be removed by baking or prolonged pumping.

2.4.8. SUMMARY

This instrument is a useful analytical tool for residual gas analysis, especially considering its cost. The instrument has a wide range of applications and the only limitation is that, although the sensitivity is adequate for most applications, the minimum partial pressure, of 2×10^{-10} torr, detectable, is not, at present, low enough to allow the instrument to be usefully operated much below a total pressure of 10^{-9} torr.

REFERENCES

1. STECKELMACHER, W., EVANS, S. and COLLINS, R. D. 'The design and evaluation of a fast response mass spectrometer and the effect of low temperature cathodes on the ion source performance.' *Int. Symp. Residual Gases in Electron Tubes (III) and Sorption–Desorption Phenomena in High Vacuum.* Rome (1967) In press
2. STECKELMACHER, W. 'The measurement of pressure in ultra-high vacuum systems'. *Trans. Vacuum Symp.* Boston (1962)
3. CRAIG, R. D. and HARDEN, E. H. *Vacuum* 16 (1966) 67

2.5

A CYCLOIDAL PATH MASS SPECTROMETER APPLIED TO THE MEASUREMENT OF THE SPEED OF SPUTTER ION PUMPS

D. Andrew

2.5.1. INTRODUCTION

THE sputter ion pump, first introduced by Hall[1] in 1958, has now become established as an important pumping device for ultra-high vacuum. Recently, various manufacturers* have introduced a number of new designs of sputter pump for which they claim various advantages. It is thus essential that a reliable method for assessing pump performance should be available.

The accepted measure of pump performance is the pump speed S, defined by the equation:

$$S = \frac{Q}{P}$$

where P is the equilibrium pressure obtained in the pump when a steady flow of gas, Q, is introduced. For diffusion pumps in particular, many workers have investigated the techniques for obtaining reliable results and the general principles of the design and operation of suitable apparatus are now well established[2].

In the case of sputter ion pumps some special problems arise. The pump speed is found to vary considerably with the condition of the pump and the nature of the gas introduced. Also the assumption that a condition of equilibrium arises when a steady flow of gas is introduced to a sputter ion pump requires some qualification. Thus, there is no single value of pump speed which can be quoted to specify pump performance without the risk of ambiguity.

Careful examination of ion pump speed measurements shows that, under conditions used in practical experiments, the observed instantaneous pump speed may vary by a factor of five. The highest values of pump speed occur after subjecting the pump to special

* General Electric triode ion pump; Mullard magnetron ion pump; Ultek differential ion pump; Varian slotted cathode, large cell ion pump.

processing techniques, known to be beneficial for the pump. Baking the pump at 250°C while the voltage is applied, or introducing argon at 10^{-5} torr, are two of the techniques commonly employed. After such treatment, the pump speed decays slowly as pumping proceeds. The rate of decay depends on the quantity of gas pumped, following the activation process. Using spectroscopically pure nitrogen as the gas supply, the decay curves obtained with four types of sputter ion pump with widely different designs of cell geometry and size, are shown in *Figure 2.5.1.* The curves obtained from the various pumps have the same characteristic shape. The curves are quite reproducible when similar conditions are employed.

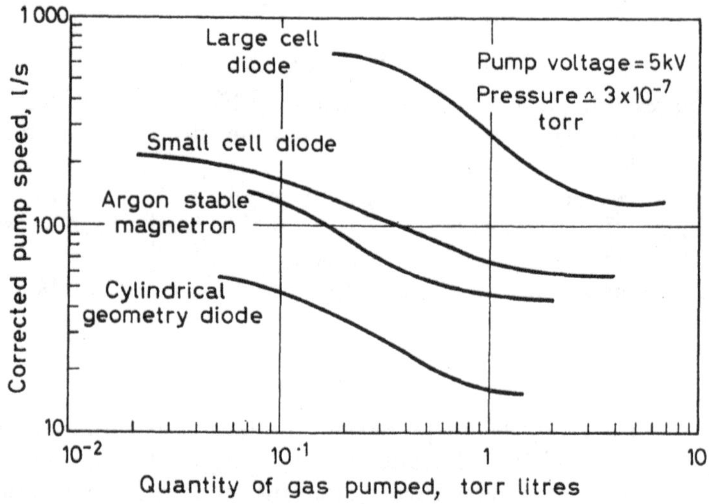

Figure 2.5.1. Variation of pump speed for nitrogen after argon treatment

It appears that a nominal 100 l/s pump requires to pump up to 5 torr litres of 'pure' nitrogen before there is even an approach to equilibrium. In order to pump this quantity of gas at a pressure of 10^{-9} torr, the pump would require a period of approximately 1½ years. This illustrates that even if an equilibrium pump speed were eventually obtained, it would be of little significance to anyone performing day-to-day experiments under conditions of ultra-high vacuum. The information contained in the curves of decay of pump speed against quantity of gas pumped, has much more application, since it will enable an operator to see the maximum pump speed to be expected after activation and to determine the rate of decay at the particular conditions under which he is operating the pump.

Further improvements to the technique of assessment are necessary, because, in practice, one pumps gas mixtures, and the presence of small quantities of inert gases, which are pumped rather slowly by sputter ion pumps, may greatly influence the pump performance. The use of mass spectrometry techniques is the next logical step. The following account describes how this technique has been applied and some results are given.

2.5.2. THE PUMP SPEED APPARATUS

The orifice flow method has been found particularly suitable for measuring the speed of sputter ion pumps. In principle, the flow of gas Q is determined by measuring the pressure drop across an

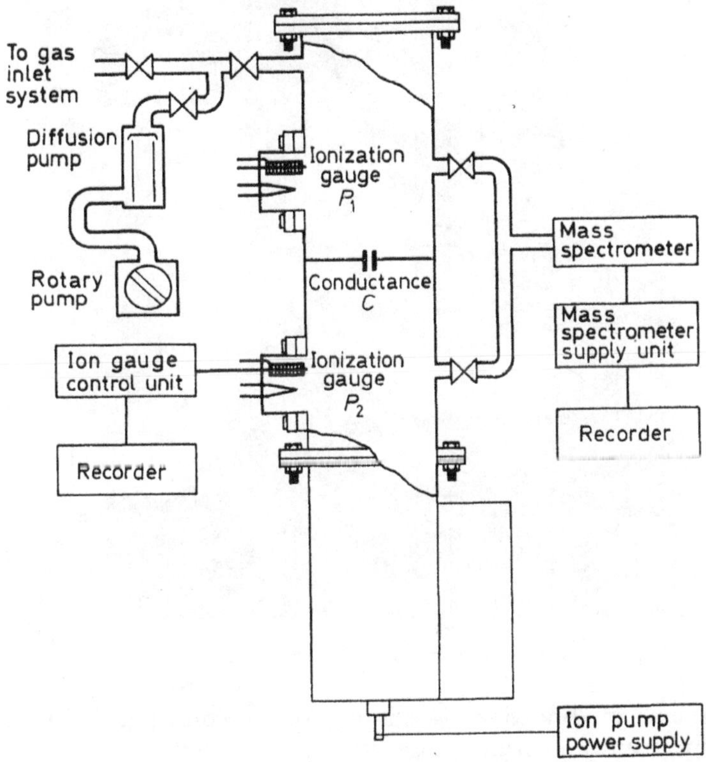

Figure 2.5.2. Apparatus for pump speed measurement

orifice of known dimensions. The pump speed is then calculated from the equation:

$$S = C\left(\frac{P_1}{P_2} - 1\right)$$

where P_1 is the reading of the Bayard–Alpert gauge on the gas inlet side of the orifice and P_2 is the pressure on the pump side of the orifice. The conductance of the orifice is calculated using the appropriate method[3]. The apparatus used in the laboratory for speed measurements has been described in an earlier paper[4].

Figure 2.5.3. The apparatus

The first modification to this apparatus consisted of placing a mass spectrometer on the pump side of the orifice.

Measurements with this apparatus showed that when the pump had pumped 5 torr litres of gas, obtained from a bottle of spectroscopically pure nitrogen, the residual gases in the pump were mainly nitrogen and argon in the ratio 10 to 1. The ratio remained practically unchanged when the pump total pressure was varied through four orders of magnitude by altering the flow of gas. It could not be proved conclusively how the high residual of argon originated in the pump.

The apparatus was, therefore, modified again and now takes the form shown schematically in *Figure 2.5.2.* The mass spectrometer was connected through two bakable valves to two new side ports, placed symmetrically with respect to the orifice, and at 90° with respect to the total pressure gauges.

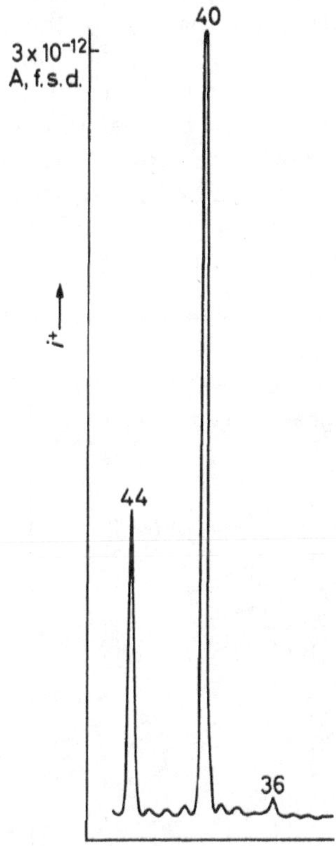

Figure 2.5.4. Mass spectrum obtained by electronic scanning of cycloidal path mass spectrometer

The mass spectrometer used is a development of the cycloidal path instrument which was described at Stuttgart[5]. It has been provided with a permanent magnet, weighing 6·5 kg, which allows the mass range from 2 u to 50 u to be obtained by scanning electrically in one sweep. The mass spectrometer is shown attached to the apparatus in the photograph reproduced in *Figure 2.5.3.*

A portion of a mass spectrum, obtained with this magnet arrangement, is reproduced in *Figure 2.5.4.* The resolution is such that neighbouring peaks at mass 40 are completely resolved, even when the peak heights differ by a factor of 100. With this adjustment the sensitivity for nitrogen 28 is 3×10^{-5} A/torr with an emission current of 100 μA.

2.5.3. EXPERIMENTAL TECHNIQUE

The apparatus was prepared for speed measurements by baking at temperatures between 300° and 400°C while pumping with the trapped diffusion pump. Outgassing tests and comparison of the gauge sensitivities, were carried out. After activation of the pump by argon treatment, the pump was left to reach a pressure in the low 10^{-10} torr range. By suitable manipulation of the bakable valves, an analysis of the residual gases in both the upper and lower sections of the apparatus was carried out. The spectroscopically pure nitrogen was then introduced and the flow controlled until the pump pressure was approximately 2×10^{-7} torr.

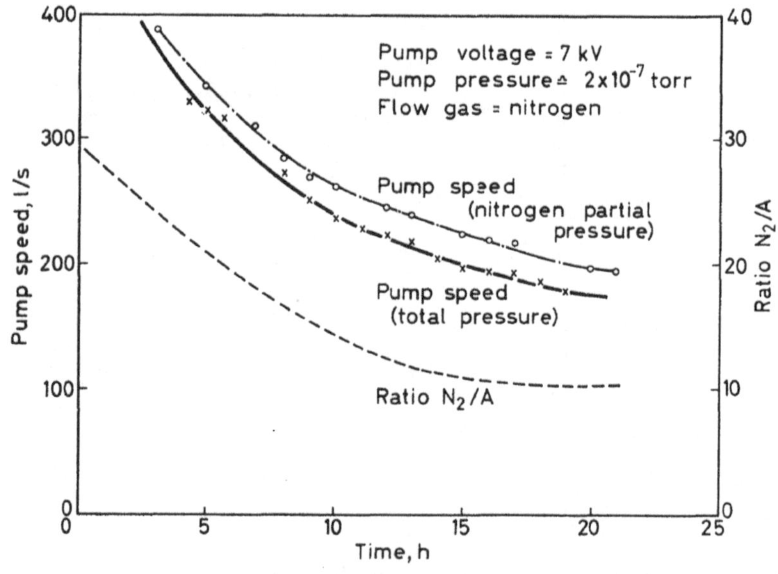

Figure 2.5.5. Decay curves for large cell diode in nitrogen

The readings of the total pressure gauges were recorded continuously while the mass spectrometer was used, to check the gas composition of both sections of the apparatus.

Analysis of the residuals on the gas inlet side of the orifice revealed that there was at least 1 part per 1000 of argon in the nitrogen in this section of the apparatus. Since the partial pressure of argon in the pump was less than that of the gas inlet, the flow was always into the pump. Thus the possibility of the pump being the source of argon is, in this case, precluded and the purity of the gas supply is suspected.

In spite of attempts to flush the gas supply lines with nitrogen, to reduce the possibility of the contamination of the gas supply by the argon originally used for processing the pump, no improvement in gas purity was obtained. Nitrogen obtained from boiling liquid nitrogen also gave similar results.

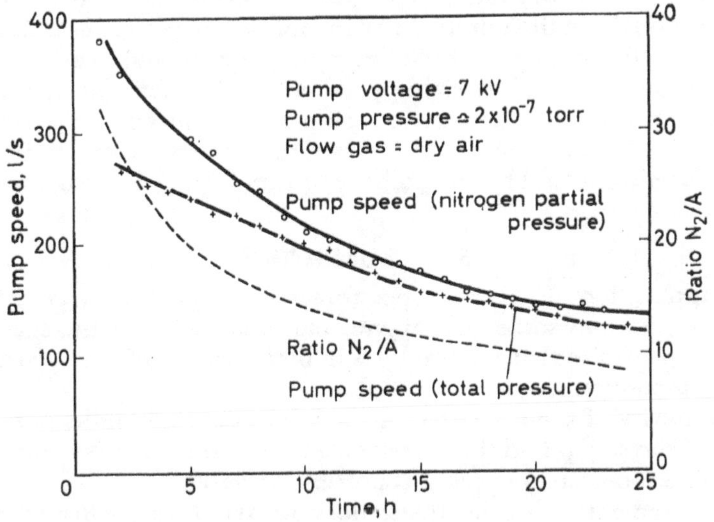

Figure 2.5.6. Decay curves for large cell diode pump in air

In view of the unexpectedly large argon content in the gas entering the pump, the question now arises whether or not the decay in pump speed observed with total pressure gauges is, in reality, due only to an accumulation of argon in the pump.

To investigate this question the mass spectrometer was set to scan from mass 26 to 46 in a period of 5 min, using the repetitive scanning facility incorporated in the control unit. Successive scans were made of the residuals in the pump during the whole period of the decay of pump speed. Intermittently, the mass spectrometer was switched to investigate the residuals in the top chamber, but no change in this gas composition was observed. From the data collected it was possible to determine the total pressure speed curve, from the Bayard–Alpert gauges, the partial pressure speed curve, from the readings of the mass spectrometer set to mass 28, and also the variation of the nitrogen and argon composition of the residuals in the pump. These results are plotted as a function of time in *Figure 2.5.5.* The pump was operated at a pressure of approximately 2×10^{-7} torr with 7 kV applied.

It is at once obvious that, in addition to a steady increase in the argon content of the gas in the pump, there is also a decline in the pumping speed for nitrogen.

Using air as the flow gas, the same method of operation yields the results plotted in *Figure 2.5.6.* Although the argon content in the air is at least ten times that of the nitrogen, the final values of the ratio N_2/Ar in the pump, differ by only a few per cent. A higher pump speed for argon, while pumping air, would lead to this observation, but the interesting possibility arises that the speed for argon depends on the mixture of gas introduced. Perhaps a similar dependence accounts for the differences observed in the speed for nitrogen. This possibility requires further investigation.

2.5.4. SUMMARY

The application of a mass spectrometer to the measurement of pump speeds has some advantages, particularly when the spectrometer is placed so that the residuals in both sections of the apparatus can be monitored.

The purity of the gas entering the pump certainly influences the speed of the pump and the presence of a mass spectrometer provides a suitable monitor to check this possible variable.

Measurements with the mass spectrometer have confirmed that the condition of the pump continually changes throughout the pumping process, so that no equilibrium condition arises, within time periods of interest to practical workers.

The decay of total pump speed is associated with a decay of the pump speed for nitrogen as well as an accumulation of argon in the system.

I am grateful to A. D. Martin who has recorded the data on which this paper is based.

REFERENCES

1. HALL, L. D. *Rev. scient. Instrum.* 29 (1958) 367
2. Conference on Performance Assessment of High Vacuum Pumps. *Vacuum* 16 Nov./Dec. (1966)
3. DUSHMAN, S. *Vacuum Technique,* 2nd edn., J. LAFFERTY (Ed.), Wiley, London, 1962
4. ANDREW, D. *Vacuum* 16 (1966) 653
5. ANDREW, D. *Trans. 3rd Int. Vacuum Congr.,* vol. 2, Pergamon, London, 1966

DISCUSSION

J. F. J. TODD (UNIVERSITY OF KENT)
Could the increase in the argon/nitrogen ratio be due to some sort of fractionation at the standard conductance?

D. ANDREW, *in reply*. This is highly unlikely. The gas passes through the conductance under molecular flow conditions.

P. G. FOX (CAMBRIDGE UNIVERSITY)
Is this argon evolution generally true for ion pumps?

D. ANDREW, *in reply*. Pumps which have pumped a few torr litres of argon generally appear to show a high argon content of the residual gas in the pump. The position is being investigated by several manufacturers using new designs of pump.

P. G. FOX (CAMBRIDGE UNIVERSITY)
Is the argon coming from welds in the system?

D. ANDREW, *in reply*. The preheat treatment given virtually ensures that all argon is removed prior to assembly.

G. F. SNOW (SCIENCE RESEARCH COUNCIL)
You did briefly look at the strange things happening directly after switching off, with respect to the mass spectra. Could you expand on this?

D. ANDREW, *in reply*. When the pump is switched off, there is always a rise of pressure of an order or two. There seems to be a trapping of gas in the pump by the potentials applied to it. As this gas is given off, it gives rise to an increasing unknown background, which would interfere with a direct analysis of the gas introduced from the gas supply.

2.6

THE SORPTION OF GASES BY THIN FILMS

D. S. Allam

2.6.1. TYPES OF SORPTION

THERE are three distinct processes which may occur between a gas and a surface. In the first type, called physical adsorption or Van der Waals' adsorption, the adsorbed gas molecules are held on the surface by weak forces of the type responsible for cohesion in liquids. In the second type, called chemical adsorption, there is a reaction between the gas molecules and the surface, leading to a satisfaction of the surface valency states. The third type is absorption or occlusion, where gas molecules occupy interstitial states in the bulk of a solid.

A gas molecule, physically adsorbed on a surface, possesses a certain mobility on the surface. Chemically adsorbed molecules are

Table 2.6.1. Classification of Metals and Semi-metals Based on Adsorption Properties from G. C. Bond[1], by courtesy of Academic Press

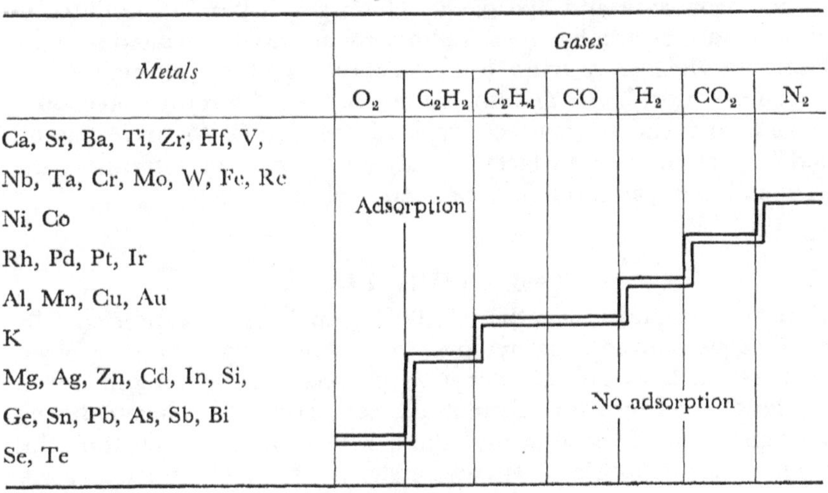

less mobile and a higher activation energy is required for movement across a surface than in the case of physical adsorption. The activation energy for desorption is also higher for chemical adsorption than for physical adsorption.

Table 2.6.1 shows a classification of metals in thin-film form, based on their ability to chemically adsorb various gases. Oxygen is adsorbed by all metals except gold. In general, metals with unfilled *d* orbitals, i.e. the transition metals, have a high affinity for most gases. It seems that chemical adsorption involves a type of covalent bond between the surface and the gas molecule. There is evidence that some surfaces act as catalysts for gaseous reactions by bringing gaseous reactants together on adjacent surface sites with the formation of a transition state, followed by desorption of the reaction product.

The surfaces in all vacuum systems contain some adsorbed gas; the pressure in a leak-free system reaches an equilibrium between the outgassing of the interior and the pumping action of the pump.

2.6.2. THE SOURCE OF GASES IN A VACUUM SYSTEM

The main sources of gas are from the walls and interior of the vacuum chamber, which may be made of glass or steel, from materials used in vacuum seals such as rubber *O* rings and from the pump itself, especially in the case of rotary and oil diffusion pumps and mercury diffusion pumps. The most common residual gases, excluding pump oils, are water vapour, oxygen, nitrogen, carbon monoxide and hydrogen. Hydrogen is present occluded in most metals, especially the refractory metals used in heated sources. Pumping alone may remove the weakly bound gas molecules, but to achieve ultra-high vacuum (better than 10^{-8} torr) it is necessary to bake the vacuum system, typically to 350°C, to drive gas off and out of interior surfaces. Only in very clean systems is the permeation of gas through the system wall a problem.

2.6.3. THIN FILMS

Thin films (typically less than 10,000 Å thick) may be prepared by vacuum evaporation, or by sputtering in a partial pressure of an inert gas such as argon. Sorption processes involving thin films, can be divided into two distinct classes; those which occur during the deposition of the film and those which occur on the thin-film surface after deposition. It can be shown[2] from the kinetic theory of gases, that the number of gas molecules (N_1) striking unit area of a surface per second is given by the following relationship:

$$N_1 = 3{\cdot}51 \times 10^{22}\, P(M_1 T)^{-\frac{1}{2}}$$

where M_1 is the molecular weight of the gas, T is its absolute temperature and P is the pressure in torr.

Not all the gas molecules striking a clean surface will adhere, the sticking probability is often less than unity. However, at a pressure of 10^{-6} torr, a clean surface would become completely covered by adsorbed gas molecules in a few seconds. *Figure 2.6.1* shows the relationship between the pressure and the time for a surface to become covered with a monolayer of adsorbed gas. Two values of the sticking probability β are given. At 10^{-6} torr a clean surface is contaminated in minutes, while at 10^{-10} torr, contamination takes several hours.

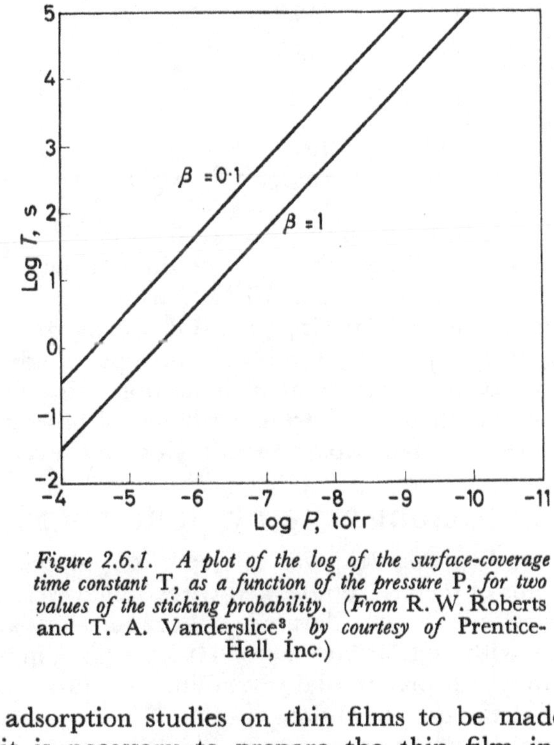

Figure 2.6.1. A plot of the log of the surface-coverage time constant T, *as a function of the pressure* P, *for two values of the sticking probability.* (*From* R. W. Roberts and T. A. Vanderslice[8], *by courtesy of* Prentice-Hall, Inc.)

For gas adsorption studies on thin films to be made with any precision, it is necessary to prepare the thin film in ultra-high vacuum conditions and maintain the low pressure before allowing into the vacuum system the pure gases to be studied. The number of molecules (N_2) of film material deposited per unit surface area per second, is given by the relationship:

$$N_2 = \frac{N_A \rho}{M} \times 10^8 \times \text{rate of deposition (Å s}^{-1}\text{)}$$

where N_A is Avogadro's number and M and ρ are the molecular weight and density of the film material, respectively. The ratio $N_1/N_2 = K$ is the number of gas molecules per film molecule arriving at the surface. *Table 2.6.2* shows values of K for tin at

Table 2.6.2. Value of K for Tin Deposited at Room Temperature. (*From* H. L. Caswell[4], *by courtesy of* Academic Press.)

P_{O_2} (torr)	*Deposition rate* (Å s^{-1})			
	1	10	100	1000
10^{-9}	10^{-3}	10^{-4}	10^{-5}	10^{-6}
10^{-7}	10^{-1}	10^{-2}	10^{-3}	10^{-4}
10^{-5}	10	1	10^{-1}	10^{-2}
10^{-3}	1000	100	10	1

various rates of deposition, in various pressures of oxygen, assuming sticking probabilities of unity for both the tin and the oxygen.

Even at 10^{-7} torr and a rate of 10 Å s^{-1} a tin film could contain 1% of oxygen. In fact, the sticking probability of oxygen is probably less than 0·1, so the impurity levels are correspondingly lower. A tin film prepared at 10^{-9} torr at a deposition rate of 1000 Å s^{-1} would contain less than one part per million of gaseous impurity and at 10^{-9} torr its surface would remain clean for several hours.

2.6.4. GAS ADSORPTION BY PURE THIN FILMS

The pioneer work in this field was carried out by Beeck *et al.*[5]. He evaporated films of metals from heated wires which were previously carefully outgassed. Films formed by evaporation are usually polycrystalline with high surface areas. It is possible, in some cases, to prepare films with preferential crystalline orientation, by careful control of the deposition conditions.

The adsorption of gases by clean thin-film surfaces may be studied by several techniques, including electrical methods[6], infrared spectroscopy[7] and, of course, mass spectrometry. The advantages of the latter are obvious, since it is important to know which species are entering into the sorption process, especially when gaseous reactions are taking place activated by the film surface. For example, in the case of the reaction between a hydrocarbon, such as methane, and deuterium catalysed by a nickel film, measurements can be made either on the increase of deuterium in the hydrocarbon or the increase of hydrogen in the deuterium. For this type of experiment there is no substitute for mass spectrometry[8].

The reaction between methane and deuterium, for example, gives rise to the following products: CH_3D, CH_2D_2, CHD_3, HD and H_2. It is necessary to analyse the gas mixture (including the original gases CH_4 and D_2). This is relatively straightforward for the simple example of methane and deuterium. The method of analysis is briefly as follows.

When the energy of the electrons in the mass spectrometer is lower than the ionization potential of the molecule, no ions are formed. When the electron energy is significantly greater than the ionization potential, dissociation occurs. The minimum energy required to form a given ion is termed the 'appearance potential' of that ion. It would be convenient to work below the appearance potential of all fragment ions and, therefore, only observe the mass spectra of the parent ions and convert the relative ion currents into a percentage composition. In practice this is not possible. Electron energies of between 12 eV and 20 eV are normally used and these lead to fragmentation of the molecules. The analysis is quite straightforward if only a few fragment ions are formed. Complications are introduced when, in the case of some substituted hydrocarbons, two or more fragments have the same mass, e.g. $C_2H_2D_4^+$ and $C_2D_5^+$. The use of mass spectrometry does allow the identification of isomers by their different fragmentation patterns. For example, the deuterated propane $C_3H_3D_2$ may have only one of the following structures: CH_3—CD_2—CH_3, CHD_2—CH_2—CH_3, CH_2D—CHD—CH_3, CH_2D—CH_2—CH_2D.

Figure 2.6.2 shows the results of a typical experiment[9] in which a clean 150 Å thick rhodium film was deposited at 10^{-9} torr. After deposition, ethane was admitted and the composition of the gas phase monitored, using a mass spectrometer. At 27°C the only gaseous product was methane, some carbon and hydrogen produced remained adsorbed on the rhodium film. The overall reaction was found to be:

$$C_2H_6 \text{ (gas)} \rightarrow 1{\cdot}2\ CH_4 \text{ (gas)} + 0{\cdot}8\ CH_{1{\cdot}5} \text{ (residue)}$$

To investigate the nature of the residue, deuterium was allowed into the system. The reaction products were found to be CD_4, CHD_3 and CH_2D_2, indicating that only monocarbon fragments were present on the rhodium surface.

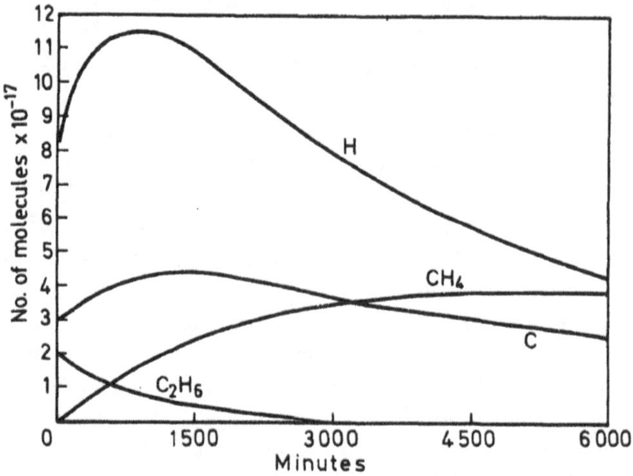

Figure 2.6.2. Rate of desorption of ethane on a 27°C *rhodium surface and the rate of product formation.* (*From* R. W. Roberts[9], *by courtesy of* Transactions of the Faraday Society)

2.6.5. ADSORPTION OF GASES DURING THE DEPOSITION OF THIN FILMS

During evaporations or sputtering, carried out in conventional high vacuum systems, the value of K is usually such that gas molecules will be adsorbed or buried by the film during deposition and will remain in the body of the film. Subsequent heat treatment processes and diffusion may alter the distribution of the gas molecules, but in many cases, reaction occurs and the gas remains as an impurity in the film. Another important factor influencing deposition is that adsorbed gases on the substrate, prior to deposition, may affect the sticking probability and adhesion of the film material.

In certain cases the adsorption of gases by thin films is used to advantage, for example in reactive sputtering or evaporation, thin films of certain oxides may be prepared by depositing the appropriate metal in a partial pressure of oxygen. The titanium sublimation pump operates by the deposition of the highly reactive metal titanium which reacts with gases to form involatile solids and at the same time buries non-reactive gases such as argon.

In some cases, the presence of impurities in thin films can have an adverse effect on certain properties of the films. A good example of this is the case of superconducting thin films, where the presence of impurities affects the position and the sharpness of the superconducting-resistive transitions. *Figure 2.6.3* shows superconducting-resistive transitions for thin films of tin[10] (the K values are given as percentages). For films prepared in high vacuum the transition is sharp; as the partial pressure of oxygen is increased, a higher magnetic field is required for a less sharp transition to occur.

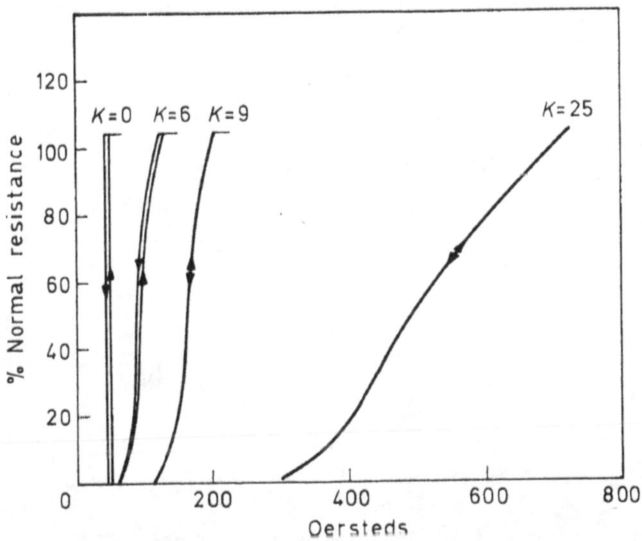

Figure 2.6.3. Resistance as a function of applied magnetic field for oxygen-doped tin films. (*From* H. L. Caswell[10], *by courtesy of* Journal of Applied Physics)

Caswell has suggested critical partial pressures of gases which should not be exceeded for the production of high quality reproducible tin films. These are 5×10^{-8} torr oxygen, 4×10^{-7} torr water vapour and 8×10^{-7} torr carbon dioxide.

Figure 2.6.4 shows the variation of some of the more important residual gases during the deposition of tin films[11]. The main features are common to many systems, hydrogen is liberated by the source material and its charge. Even if the source is carefully outgassed, more hydrogen is released when the temperature is raised to that required for evaporation. Oxygen is removed during the evaporation; the levels of carbon dioxide and water vapour increase. *Figure 2.6.5* shows the effect of oxygen on the structure of an

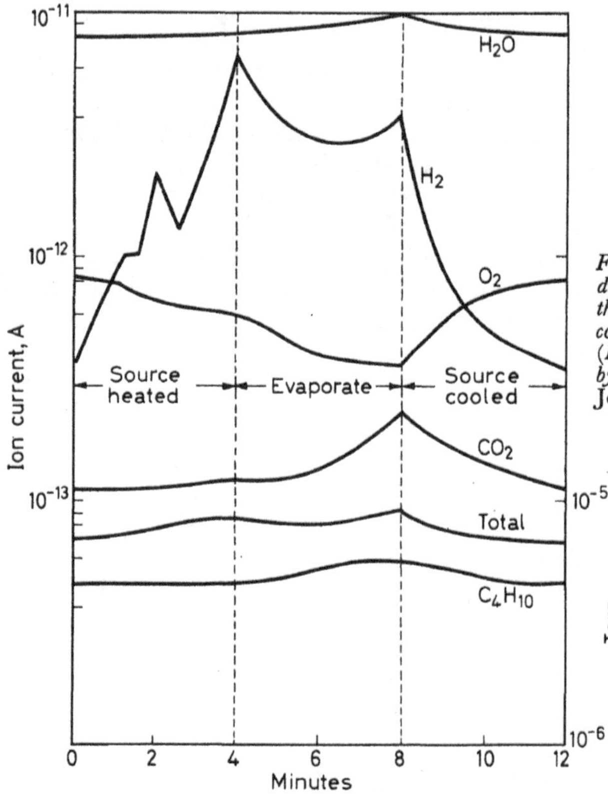

Figure 2.6.4. The residual gases present during the deposition of tin in a convential evaporation. (From H. L. Caswell[11], *by courtesy ol* IBM Journal of Research and Development)

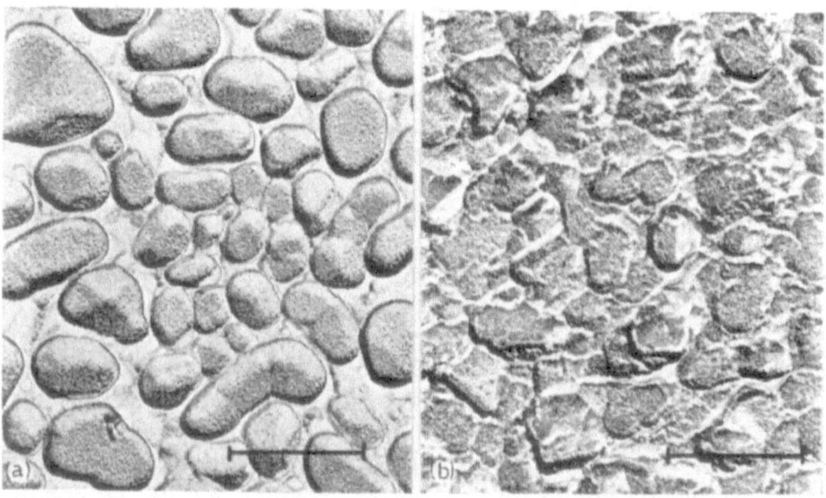

Figure 2.6.5. Replica electron micrographs of tin films evaporated at (a) 1×10^{-3} torr *and* (b) 1×10^{-5} torr oxygen; *the marker indicates 1 micron; the films were evaporated at approximately* 100 Å s^{-1} *and were about* 1000 Å *thick.* (*From* J. B. Preece, C. T. H. Stoddart and H. Wilman[12], *by courtesy of the* Philosophical Magazine)

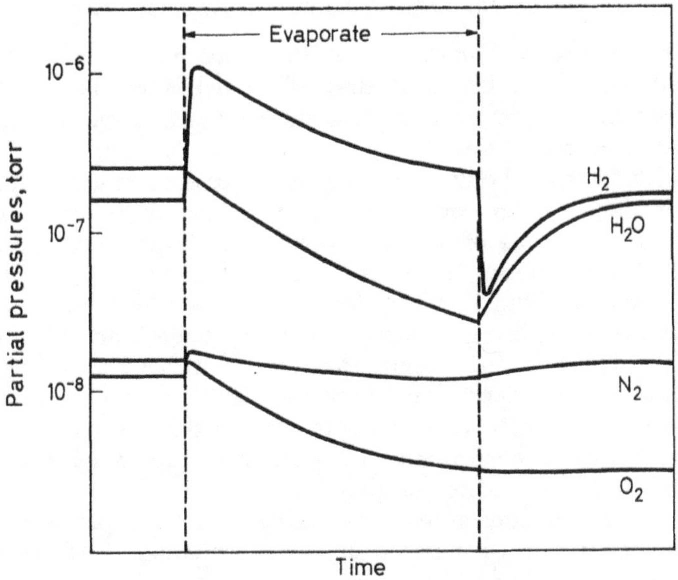

Figure 2.6.6. The residual gases present during the evaporation of silicon monoxide in a convential evaporation

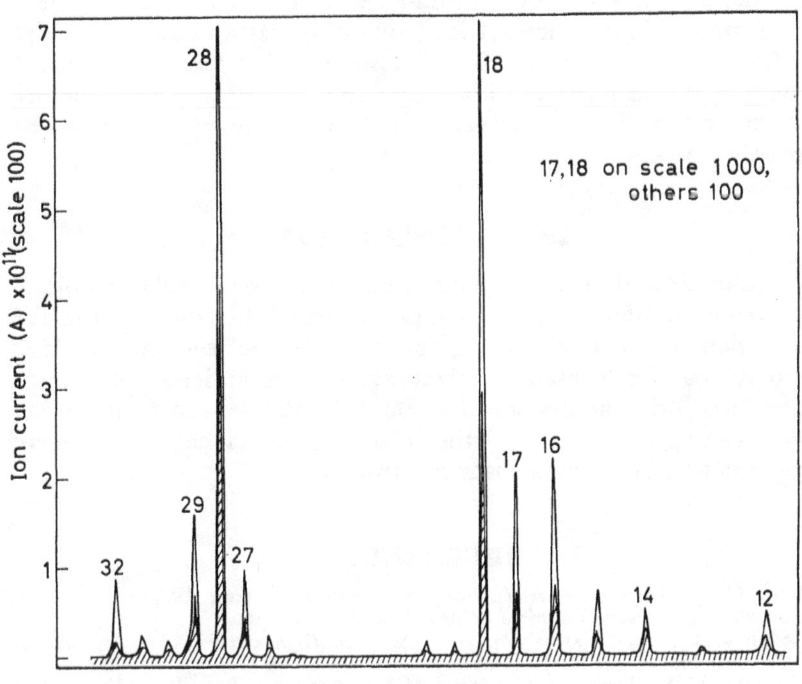

Figure 2.6.7. The mass spectra of residual gases in a vacuum system before and after evaporation of silicon monoxide (shaded spectrum after evaporation)

evaporated tin film[12]. The mobility of tin deposited at 1×10^{-8} torr [*Figure 2.6.5(a)*] is higher than that of tin deposited at 1×10^{-5} torr [*Figure 2.6.5(b)*] of oxygen, thus the former film exhibits large crystallites in its structure.

Figure 2.6.6 shows the changes of partial pressure of gases during an evaporation of silicon monoxide. In this case hydrogen is again evolved, the pressures of water vapour and oxygen are reduced during the evaporation. There is an overall decrease in the total pressure, typically from 1×10^{-6} torr to 5×10^{-7} torr. *Figure 2.6.7* shows the mass spectra before and after deposition of silicon monoxide. There is little doubt that oxidation occurs during the evaporation and condensation of the metastable silicon monoxide, both oxygen and water vapour contribute to this. Hydrogen has also been shown to be present in silicon monoxide films, both by electrical[13] and infra-red techniques[14].

Many other thin-film systems probably owe their properties to the presence of impurity gases in their structures. It is often impossible to reproduce the properties of the bulk material, for example, the resistivities of chromium films can be as high as four times the resistivity of bulk chromium metal[15].

In the manufacture of thin-film resistors based on chromium–silicon monoxide mixtures, it is common for resistivities to increase by 100% on baking in air after deposition. This has been shown[14] to be due to the partial oxidation of the imperfect silica structure and the removal of hydrogen which was adsorbed during the deposition process.

2.6.6. CONCLUSION

It is quite clear that mass spectrometry is an invaluable technique in the examination of the surface properties of thin films and in the elucidation of processes during the deposition of thin films. It is common practice to use an ionization gauge to indicate the change of pressure during an evaporation, but it is often far more interesting and meaningful to follow the changes of partial pressure of component gases, using a mass spectrometer.

REFERENCES

[1] Bond, G. C., *Catalysis by Metals*, Academic Press, New York, 1962
[2] Dushman, S., *Vacuum Technique*, Wiley, New York, 1949
[3] Roberts, R. W. and Vandersclice, T. A., *Ultra High Vacuum and its Applications*, Prentice-Hall, New York, 1963
[4] Caswell, H. L., *Physics of Thin Films*, Vol. 1, Academic Press, New York, 1963

[5] BEECK, O., SMITH, A. E. and WHEELER, A. *Proc. R. Soc.* A177 (1941) 62
[6] SUHRMANN, R. *Adv. Catalysis* 7 (1955) 303
[7] GUNDRY, P. M. and TOMPKINS, F. C. *Quart. Rev.* 14 (1960) 257
[8] WRIGHT, P. G., ASHMORE, P. G. and KEMBALL, C. *Trans. Faraday Soc.* 54 (1958) 1692
[9] ROBERTS, R. W. *Trans. Faraday Soc.* 58 (1962) 1159
[10] CASWELL, H. L. *J. appl. Phys.* 32 (1961) 105
[11] CASWELL, H. L. *IBM J. Res. Dev.* 4 (1960) 130
[12] PREECE, J. B., and STODDART, C. T. H. and WILMAN, H. *Phil. Mag.* 16 (1967) 447
[13] SIDDALL, G. *Thin Films Conf.*, Nottingham, 1967
[14] ALLAM, D. S., and PITT, K. E. G., *Solid Thin Films,* 1 (1968) 245
[15] GOULD, P. A. *Br. J. appl. Phys.* 16 (1965) 1481

DISCUSSION

C. J. POTTER (WARREN SPRINGS)
In investigating the wettability of metallic gold, I wonder whether useful results could be obtained using thin films prepared by vacuum deposition.

D. S. ALLAM, *in reply.* The micro-structure of thin-film gold is different from that of gold in bulk, therefore any results obtained should be referred to the method of preparation of the specimen.

C. J. POTTER (WARREN SPRINGS)
Are there any preferred absorption sites in thin-film gold?

D. S. ALLAM, *in reply.* Yes, there are probably preferred sites since the gold film will usually be porous.

GENERAL DISCUSSION OF SESSION 2

L. HOLLAND (CHAIRMAN)
In flavour chemistry is it possible to define a flavour from its mass spectrometric constituents?

R. SELF, *in reply*. This is a future possibility. It has been reported that the components of a complex fruit concentrate have been identified almost completely by g.c.–m.s. A similar mixture synthesized from the m.s. data was submitted along with the original mixture to tasting specialists and similar flavours were reported after slight adjustments were made to the synthesized mixture. In general, however, this is not yet possible.

D. ANDREW (MULLARD)
Are the characteristics of multipliers stable with time in the presence of some of the peculiar substances which are used with mass spectrometers?

M. A. VETSCH, *in reply*. The life of a multiplier depends on conditions and multipliers used with g.c. outputs deteriorate more rapidly.

S. EVANS, *in reply*. In our experience Be–Cu multipliers deteriorate very rapidly but Ag–Mg remain stable generally. If, however, the latter is seriously contaminated it can be regenerated.

G. W. BALL (TWENTIETH CENTURY)
How are Ag–Mg multipliers regenerated?

S. EVANS, *in reply*. Regenerated by r.f. discharge in oxygen.

R. M. ELLIOTT (AEI)
In defence of the Be–Cu multiplier, in my experience these tend to settle down to give stable response over a long period after a rapid initial deterioration. This position is regarded as satisfactory, if the initial gain is sufficiently high.

SESSION 3

CHEMICAL APPLICATIONS I

3.1

THE USE OF A QUADRUPOLE MASS FILTER IN THE STUDY OF A REACTING SURFACE

P. G. Fox and R. F. Walker

3.1.1. INTRODUCTION

The much-improved sensitivity provided by the introduction of the experimental techniques of ultra-high vacuum and partial pressure analysis, has made possible the investigation of certain problems which hitherto have been inaccessible to experimental study. For example, two such areas of research with which we have been concerned are the interaction of gases with surfaces in the very low coverage region[1] and the mechanism of decomposition of specific crystallographic faces of some metallic azides[2].

The metallic azides decompose according to the scheme:

$$MN_3 \rightarrow M + 3/2N_2$$

where nitrogen is the only gaseous product evolved in the reaction. Recently, it has been shown that the rate of decomposition of the series M = Na^+, Tl^+ and Pb^{++}, is directly proportional to the surface area of the reacting crystal and that the form of the rate curves suggests that defects in the crystal play an important part in the decomposition process[2]. This is supported by direct observation of decomposing crystals in the electron microscope. However, the gaseous decomposition products are found to be more complex than is suggested by the above scheme.

In this paper we describe some experiments in which a quadrupole mass filter is used to follow changes in the composition of the gas phase during the slow thermal decomposition of single crystals of the above series of azides and this is related to the kinetics of the reaction. The sensitivity of the method is such that the decomposition of a monolayer of molecules from a surface 10^{-3} cm^2 can be followed.

3.1.2. EXPERIMENTAL

The apparatus and its mode of operation have been described elsewhere[2]. The all metal ultra-high vacuum system is pumped by an ion pump and a Varian gas analyser of the quadrupole mass filter type is used to follow changes in the composition of reaction products during decomposition (*Figure 3.1.1*). Throughout an experiment the system was isolated from the pump and the B.A. gauge run at 40 μA filament emission to minimize gauge pumping. Similarly this effect was reduced in the mass filter ion source by operating the filament at an emission of 100 μA.

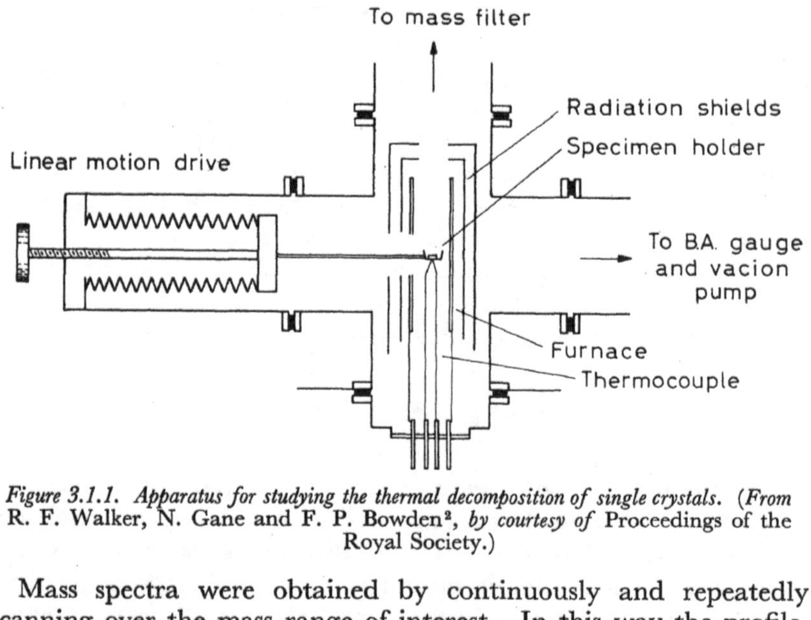

Figure 3.1.1. Apparatus for studying the thermal decomposition of single crystals. (*From* R. F. Walker, N. Gane and F. P. Bowden[2], *by courtesy of* Proceedings of the Royal Society.)

Mass spectra were obtained by continuously and repeatedly scanning over the mass range of interest. In this way the profile, formed by the peaks of each species present, gives the pressure–time curve of that species.

For all three azides, experiments were carried out on individual single crystals grown from aqueous solution.

3.1.3. RESULTS AND DISCUSSION

The most remarkable feature in the comparative behaviour of the three materials is the increasing complexity of the composition of the gas phase in the series lead → thallous → sodium azides. Moreover, this can be related to greater complexity in the pressure–time curves. Thus, apart from the initial appearance of CO_2 due to the decomposition of surface carbonate, the decomposition of lead azide is not accompanied by the evolution of dissolved gases and is also not preceeded by an induction period. This is true to the lowest decomposition temperature studied (180°C). At the other extreme is sodium azide which always shows an induction period, the extent of which decreases with increasing temperature and whose decomposition is always accompanied by the evolution of significant amounts of species other than nitrogen, e.g. the rare gases. Although the crystals were grown from aqueous solution, water does not appear in the mass spectrum, presumably due to reaction with the sodium produced in the reaction. The decomposition of thallous azide shows intermediate behaviour in which small amounts of rare gases, carbon dioxide and water, are present in the gas phase near the start of the reaction but nitrogen quickly predominates. Moreover, the induction period is observed only at the lower temperatures studied. These effects are shown in *Figures 3.1.2, 3.1.3, 3.1.4* and *3.1.5.*

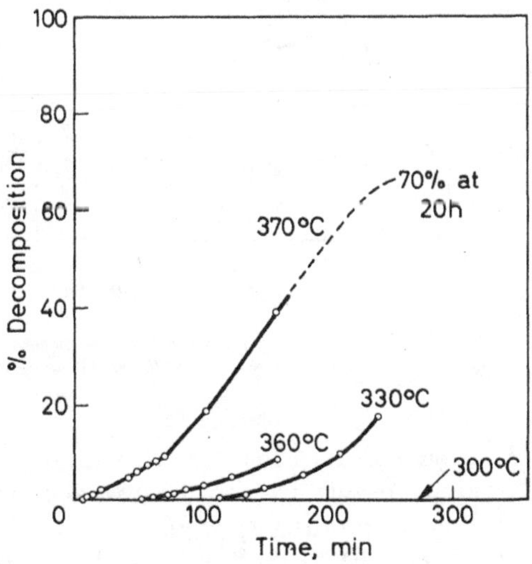

Figure 3.1.2. Kinetics of the slow thermal decomposition of sodium azide (linear plot)

The decomposition of sodium azide is plotted on a linear scale in *Figure 3.1.2* and on a logarithmic scale in *Figure 3.1.3* to emphasize the nature of the induction period. During this latter phase of the reaction, only a small percentage of the decomposition takes place and comparable amounts of nitrogen and rare gases are evolved before the reaction slows down. During the main stage of the reaction, nitrogen accounts for the majority of the gas phase and the concentration of gaseous impurities tends towards a constant value.

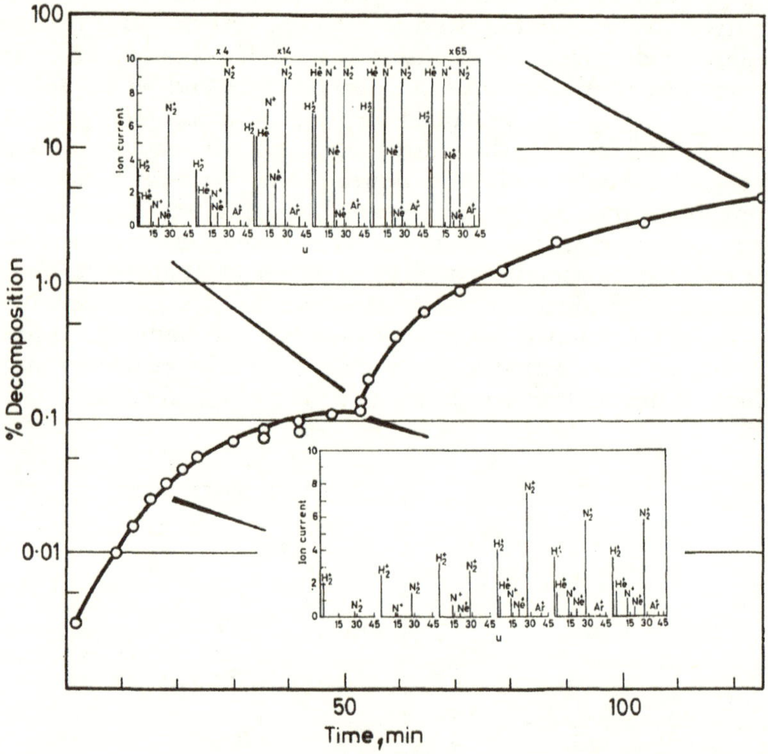

Figure 3.1.3. Kinetics of the slow thermal decomposition of sodium azide at 360°C *(log plot); the mass spectra in the insets were recorded at the same sensitivity*

In *Figure 3.1.4*, for the decomposition of thallous azide, the induction periods are shorter and accompanied by a much smaller evolution of impurity species.

In *Figure 3.1.5*, carbon dioxide from the decomposition of a small amount of basic lead carbonate on the surface of the lead azide crystal, is the only impurity recorded and this reaction is over before an appreciable amount of the azide has decomposed. The maximum amount of this surface contamination corresponds to about a monolayer in these experiments.

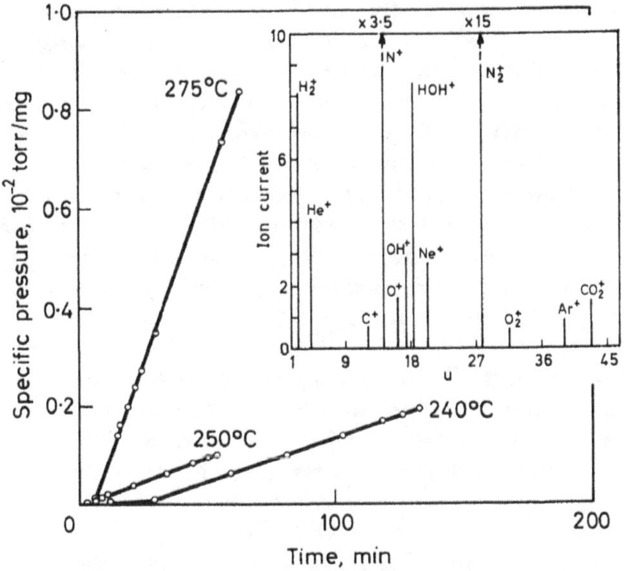

Figure 3.1.4. Kinetics of the slow thermal decomposition of thallous azide; the inset shows the composition of the gas phase near the origin of the graph

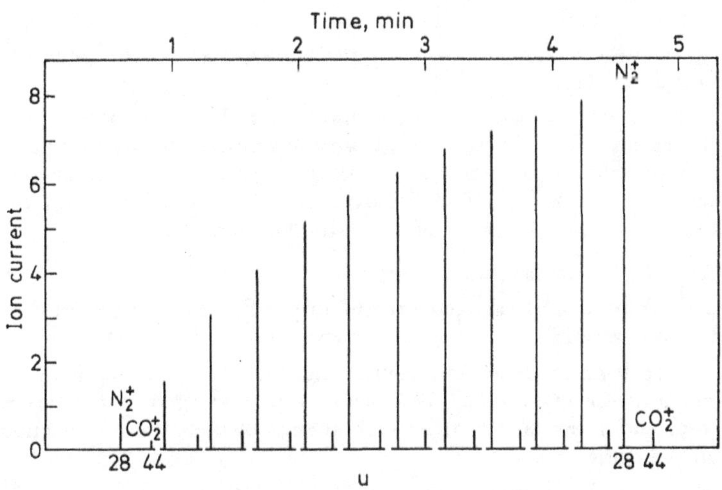

Figure 3.1.5. The kinetics of the slow thermal decomposition of lead azide at 200°C; the small amount of carbon dioxide arises from a surface contamination of basic lead carbonate

The traditional view of solid state reactions which are preceeded by an induction period, is based on the idea that, during this time, the nuclei of solid-product phase establish an interface at which the reaction can more readily take place. This is supported by a good deal of experimental evidence[3]. However, the present results suggest that a small amount of reaction taking place at impurity centres is an important characteristic of the induction period and that nuclei play a much less important rôle. This is supported by lack of any evidence for the existence of nuclei in the early stages of decomposition of all three azides as judged by observation of the reacting crystals in the scanning electron microscope at magnifications up to 30,000.

Finally, we would like to re-emphasize the greater degree of perception which is possible, in surface problems of this type, by making use of the much-increased sensitivity provided by ultra-high vacuum techniques and partial pressure measurements.

REFERENCES

[1] BAKER, B. G. and FOX, P. G. *Trans. Faraday Soc.* 61 (1965) 2001

[2] WALKER, R. F., GANE, N. and BOWDEN, F. P. *Proc. R. Soc.* 294 (1966) 417

[3] YOUNG, D. A., *International Encyclopedia of Physical Chemistry and Chemical Physics*, Vol. 1, Topic 21, Pergamon, Oxford, 1966

DISCUSSION

J. H. HUNT (ALLEN AND HANBURY)

In the preparation of azides which gave rare gases, at what stage were these incorporated?

Reply. This is not known, but we suggest that these originate from the atmosphere and are being absorbed from the solution during preparation. The amounts calculated from the solubilities of rare gases in water are the basis for this conclusion. The removal of these impurities from active centres is presumably the cause of the induction period.

D. S. ALLAM (UNIVERSITY OF SURREY)

Do you think that the hydrogen present in the first sample would depress the induction period?

Reply. There is no evidence for reaction between hydrogen and the other components of the gas phase. The presence of hydrogen in the system is an artefact and arises largely from components of the vacuum system and not from the azide crystals.

G. F. SNOW (SCIENCE RESEARCH COUNCIL)

Was the basic lead carbonate deliberately stabilized or was its presence accidental?

Reply. The presence of basic lead carbonate is accidental and is produced by the slow and almost unavoidable reaction between the surface of the azide crystals, moisture and small residual amounts of carbon dioxide in the solution from which the azide crystals were formed. Basic lead carbonate is, however, more reactive than the azide itself, so that decomposition of the former is complete before an appreciable proportion of the azide has reacted (*Figure 3.1.5*).

J. F. J. Todd (University of Kent)

Have you tried artificially increasing the rare gas concentration, to see if the induction period is increased, and does any one rare gas increase the period more than any other?

Reply. We have not done any experiments to test this possibility. Such experiments would be difficult to perform. At the moment, the concentrations of rare gas impurities are set by their rather small solubilities in the solutions from which the azide crystals were prepared.

3.2

MASS-SPECTROMETRIC INVESTIGATION OF THE FORMATION OF DI-IMIDE BY THE CATALYTIC DECOMPOSITION OF HYDRAZINE AT LOW PRESSURES ON PLATINUM

E. M. A. Willhoft and A. J. B. Robertson

A STUDY of the catalytic decomposition of anhydrous hydrazine on platinum was originally made in order to examine the effects of flashing the catalyst, which was in the form of a spiralled wire, on the decomposition reaction. Flashing a wire simply means electrically heating it for a few seconds in high vacuum, at a temperature just below the melting point of the metal. Although the gas-phase decomposition of hydrazine has received considerable attention in recent years, the heterogeneous decomposition reaction has been relatively neglected. In the literature there are no data describing the nature of the initial products formed by the decomposition of hydrazine, at low pressure, on metals. In order to identify any reactive intermediates, such as free radicals, formed as primary products after collision of molecules with the heated platinum catalyst, it is clearly necessary to work at low reactant pressures so as to minimize gas phase collisions.

The preliminary study was carried out without a mass spectrometer, using a conventional high vacuum flow system constructed of Pyrex glass. Hydrazine was admitted, at a known rate, into the reaction vessel containing the electrically-heated platinum wire. Undecomposed molecules and any condensible products were removed continuously from the vessel by condensing in a liquid air trap. Hydrogen and nitrogen produced were swept out with a mercury diffusion pump backing the liquid air trap. The number of molecules of hydrazine decomposed per second, N_s, was estimated by the difference between the number collected after each run and the known number which passed into the reaction vessel per second, N_e. It can be shown[1] that the probability, $\bar{P}$, of the reactant molecule decomposing on colliding with the catalytic wire of surface area A cm^2, is given by:

$$\bar{P} = 1{\cdot}61 \times 10^{-2} N_s S/[(N_e - N_s)A]$$

where S is the pumping speed, for unit molecular weight gas, in litres per second, of the liquid air trap at the reaction vessel. Thus a reaction probability of unity implies that every molecule colliding with the wire catalyst is decomposed.

In order to obtain further insight into the mechanism of decomposition of hydrazine, the ratio of hydrogen to nitrogen (H/N) was measured experimentally by catalytically oxidizing the hydrogen over a platinum catalyst in the backing volume, in the presence of nitrogen also collected there, and measuring the reduction in pressure with a McLeod gauge. It was also possible to determine the ratio of the amount of hydrazine decomposed to the amount of ammonia formed and collected (N_2H_4/NH_3) after a run. From earlier work[1] on ammonia decomposition under these conditions, it can be shown that the number of molecules of product ammonia decomposed on the heated filament, as a result of successive collisions with it, is negligible in relation to the number of hydrazine molecules decomposed. The leak rate of hydrazine into the reaction vessel was kept constant and it was possible to use two pumping speeds corresponding to equilibrium reactant pressures of 1×10^{-4} torr and 2×10^{-3} torr in the absence of reaction. Some results are summarized in *Figure 3.2.1* where the upper curve of each couple corresponds to the fast pumping speed conditions.

The actual speeds for condensible, S_c, and non-condensible, S_n, gases, are given in *Figure 3.2.1(a)* and refer to unit molecular weight gases. The points to be emphasized for the fast pumping speed are:

(*i*) More than three times as many molecules of hydrazine are decomposed than the calculated number corresponding to the collision rate with the wire at 950° K.

(*ii*) A considerable amount of ammonia is formed at temperatures approaching red heat. An N_2H_4/NH_3 ratio of unity is obtained at temperatures approaching 900°K.

(*iii*) Initial H/N ratios as high as 2·5 at 1250°K are obtained. After almost 2 min of reaction, this ratio decreased to a steady value (1·9 at 1250°K). The time taken to reach a steady condition is found to be independent of temperature in the range 700–1700°K. The H/N versus temperature curves corresponding to 15 s and 5 min periods of reaction prior to collecting the non-condensible products, are shown in *Figure 3.2.2.* Results described in (*i*) and (*ii*) were obtained from runs lasting either 30 min or 60 min.

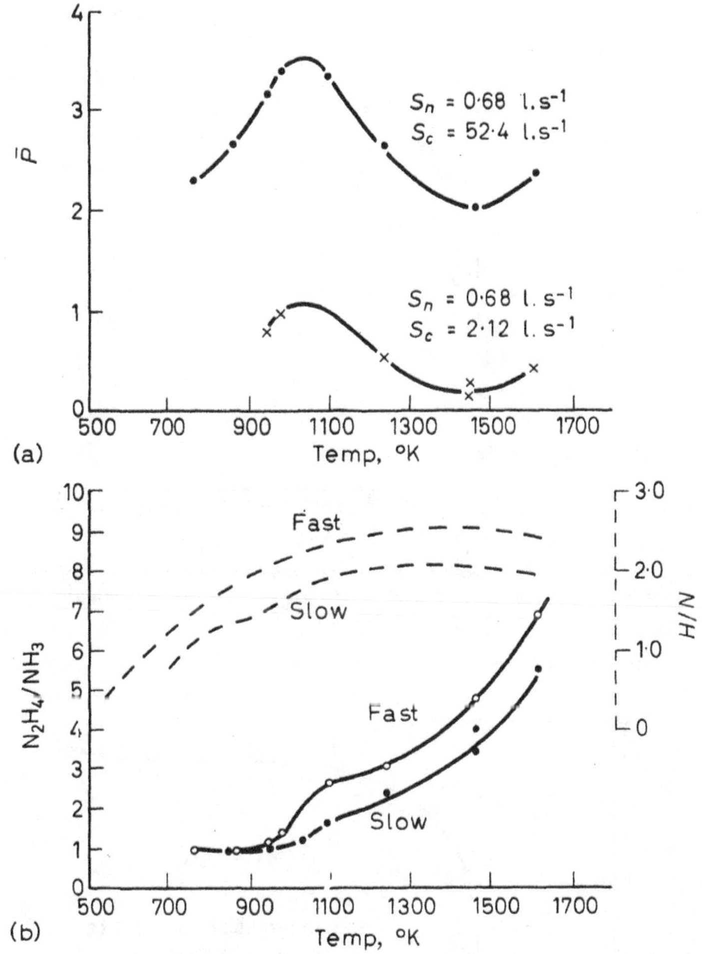

Figure 3.2.1. Kinetics and products analysis of the catalytic decomposition of hydrazine on a heated platinum wire; (a) reaction probability versus catalyst temperature for the two pumping speeds indicated; (b) N_2H_4/NH_3 *(left-hand ordinate) and initial* H/N *(discontinuous ordinate referring to discontinuous curves) ratios versus temperature for fast and slow pumping speeds*

These observations all indicate the existence of a platinum-catalysed process leading to the formation of a reactive intermediate.

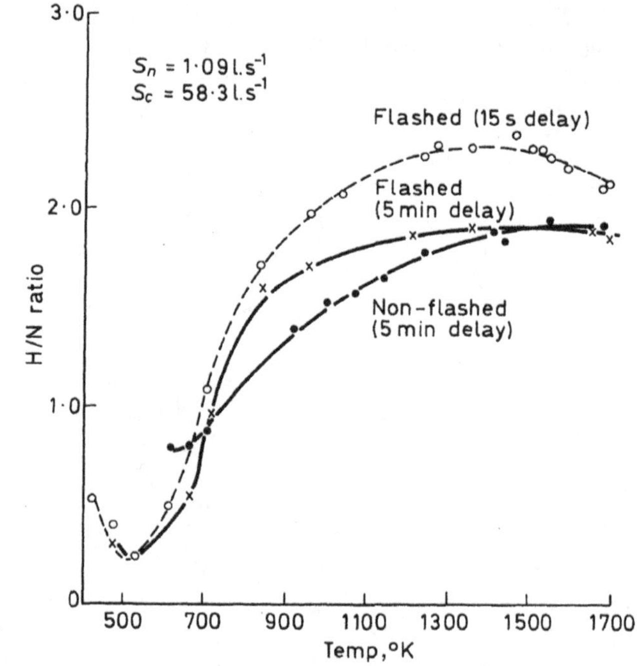

Figure 3.2.2. H/N *ratios versus catalytic wire temperature after* 15 s *and* 5 min *of reaction; lowest curve corresponds to ratios obtained for non-flashed wire*

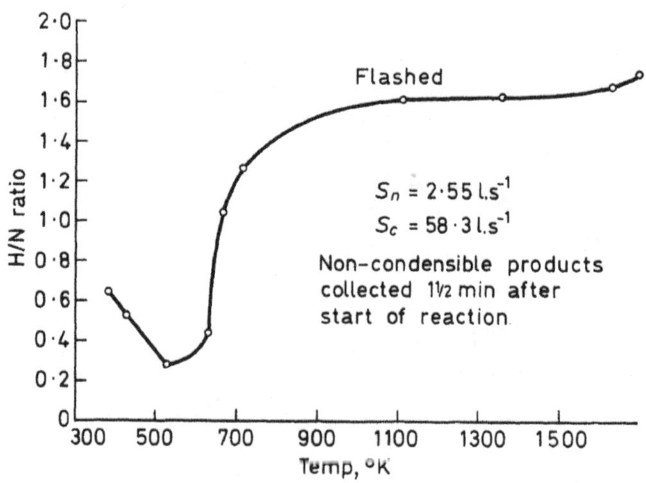

Figure 3.2.3. H/N *ratio versus catalytic wire temperature for hydrazine containing about* 2% *water*

This intermediate, or its immediate decomposition product, is able to react with both hydrazine and ammonia at the walls of the reaction vessel, giving rise to free radicals so that, in effect, there is a propagating mechanism for reactive intermediates at the walls.

The addition of about 2% water to hydrazine not only had the effect of reducing the yield of reactive intermediates (e.g. $\bar{P}$ was reduced by a factor of ten at 980°K), but also reduced the H/N ratios as illustrated in *Figure 3.2.3* which should be compared with the curve given in *Figure 3.2.2*.

A quadrupole mass spectrometer was used in an attempt to elucidate the nature of the primary step leading to the formation of the initial reactive intermediate at elevated temperatures. A similar flow apparatus, shown in *Figure 3.2.4*, was used.

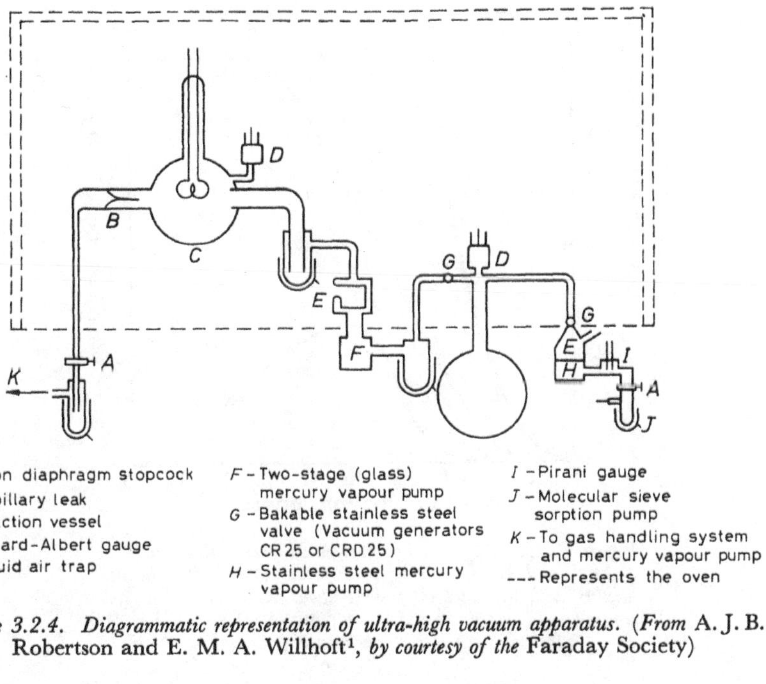

Figure 3.2.4. Diagrammatic representation of ultra-high vacuum apparatus. (*From* A. J. B. Robertson and E. M. A. Willhoft[1], *by courtesy of the* Faraday Society)

The essential difference, between this apparatus and the earlier one, is that a better vacuum could be obtained as a result of partially baking the apparatus (e.g. base pressures less than 2×10^{-7} torr). A nude ion-source quadrupole mass filter (EAI Centronic 250) was attached to the reaction vessel, so that particles could pass directly from the heated catalyst surface and into the quadrupoles.

The region surrounding the ion source was pumped separately. In order to increase, in the ion source of the spectrometer, the concentration of transient particles which evaporate directly from the platinum catalyst, the catalyst was also used in the form of a thin foil, 8 μm thick and about 2 cm by 2 cm, in a plane normal to the axis of the quadrupoles. The pressure of hydrazine for this investigation was 2×10^{-5} torr in the absence of reaction.

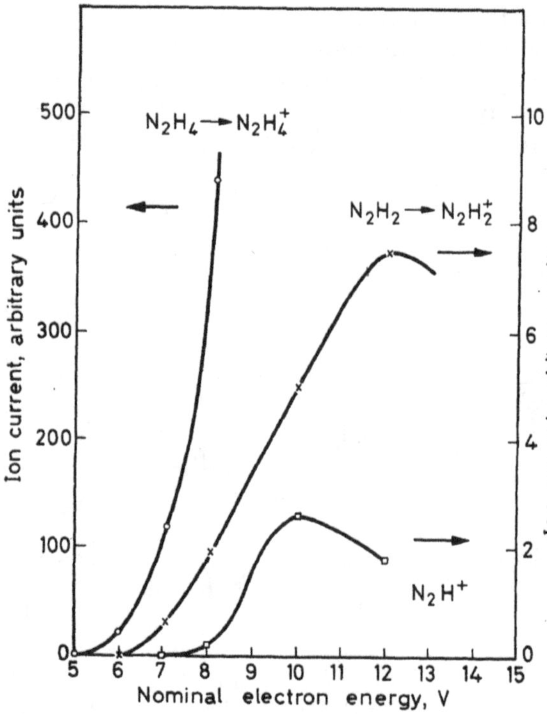

Figure 3.2.5. Approximate ionization potential of di-imide, $N_2H_2^+$ *and* N_2H^+ *curves correspond to difference in peak heights with the catalytic foil at about* 1850°K *and at room temperature; corrected ionization potential of di-imide* = 9·9 eV; *similarly, corrected appearance potential of* N_2H = 11·1 eV

On heating the foil, large relative increases in peak heights occurred, corresponding to particles of masses 30 (N_2H_2), 29 (N_2H), 16 (NH_2) and 15 (NH), at low electron energies. There was no indication of the presence of N_2H_3 and particles of mass greater than that of N_2H_4. The greatest contribution to the partial pressures of unstable particles occurred from masses 30 and 29. *Figure 3.2.5* includes curves, for the formation of $N_2H_2^+$ and N_2H^+, obtained by taking the difference in peak heights, with the foil at about 1850°K and at room temperature, for various nominal electron energies. From a literature value[2] for the first ionization potential of N_2H_4 (9·00 eV), the true ionization potential of di-imide was calculated to be 9·9 eV. Although this value is only approximate, it is very close to the value of 9·85 eV given by Foner and Hudson[3].

When ammonia alone was passed over the heated catalyst, formation of NH and NH_2 was not observed with the quadrupole mass filter. The source of the radical ions N_2H^+, NH_2^+ and NH^+ is at present conjectural since their presence may be due to any, or all, of the following four processes:

(*i*) Dissociative ionization of the primary catalytic product N_2H_2.

(*ii*) Formation as neutral chain carriers during reactions involving N_2H_4 and NH_3 with N_2H_2 or with the spontaneous decomposition product of N_2H_2, followed by ionization in the ion source.

(*iii*) Free radical decomposition of N_2H_2, either spontaneously or as a result of subsequent collisions with the platinum catalyst, followed by ionization in the ion source.

(*iv*) Additional and simultaneous modes of decomposition of hydrazine on the platinum catalyst, followed by ionization in the ion source.

The relative contributions of these various processes require further study. This might involve determining the dependencies of the various peak heights of the free radicals on the hydrazine pressure and decreasing the distance between the catalyst and the quadrupoles.

The mass spectrum of hydrazine at a pressure of 2×10^{-5} torr is given in *Table 3.2.1.*

Table 3.2.1. Mass Spectrum of Hydrazine in a Quadrupole Spectrometer with a Nominal Electron Energy of 70 eV

Mass (m/e)	14	15	16	17	28	29	30	31	32
Relative intensity	1	2	24	40	24	31	41	43	100

On heating the catalyst to 950°K from room temperature, the peak heights corresponding to masses 31 and 32 both decreased to about 62% of their values, whereas those of 30 and 29 decreased, respectively, to 70% and 76% of their previous values at room temperature. The peak height of mass 28 increased by a factor of about five. No peaks appeared with mass greater than that of hydrazine.

The simplest interpretation is that the initial reactive intermediate is di-imide, which has been reported by Foner and Hudson[4], to be formed by the decomposition of hydrazine in the gas phase and also by applying an electrodeless discharge to hydrazine. Assuming

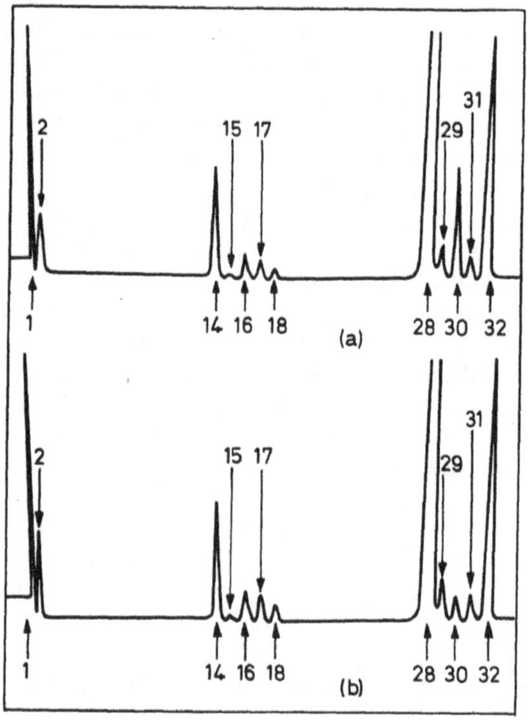

Figure 3.2.6. Mass spectra of hydrazine in the presence of air showing the effect of heating the catalytic wire in an air/N_2H_4 mixture; (a) filament at about 950°K*; (b) filament at room temperature*

equal cross sections for ionization of N_2H_4 to $N_2H_4^+$ and N_2H_2 to $N_2H_2^+$, the partial pressure of N_2H_2 was calculated from the preceeding results to be 7×10^{-7} torr. This corresponds to a reaction probability of about 0·2 for di-imide formation as a result of N_2H_4 colliding with the platinum catalyst at 950°K. Since N_2H_2 may react further at the walls of the vessel, and its partial pressure corresponds to a steady state condition, the computed value of $\bar{P}$ is low. The true value of $\bar{P}$ for the formation of N_2H_2 on collision of N_2H_4 with the catalyst, may thus be considerably closer to unity.

If the extra contribution to peak 29, when hydrazine is decomposed, results only from dissociative ionization of N_2H_2 to N_2H^+, then the relative intensities of peaks 30 and 29, in the mass spectrum of di-imide at 74 eV, would be 76 and 100 respectively. Similarly, at 16 eV the relative contributions of $N_2H_2^+$, N_2H^+, NH_2^+ and NH^+ to the di-imide spectrum would be 100, 28, 4 and 1 respectively.

When air was mixed with the hydrazine, heating the catalyst produced an increase of mass 30 by a factor of four (nominal electron energy 44 eV). This is illustrated in the mass spectra tracings shown in *Figure 3.2.6*. The graph in *Figure 3.2.6* (*b*) refers to the wire at room temperature and in *Figure 3.2.6* (*a*) to the wire at about 950°K. It should be noted that there is no corresponding increase in the peak height of mass 29.

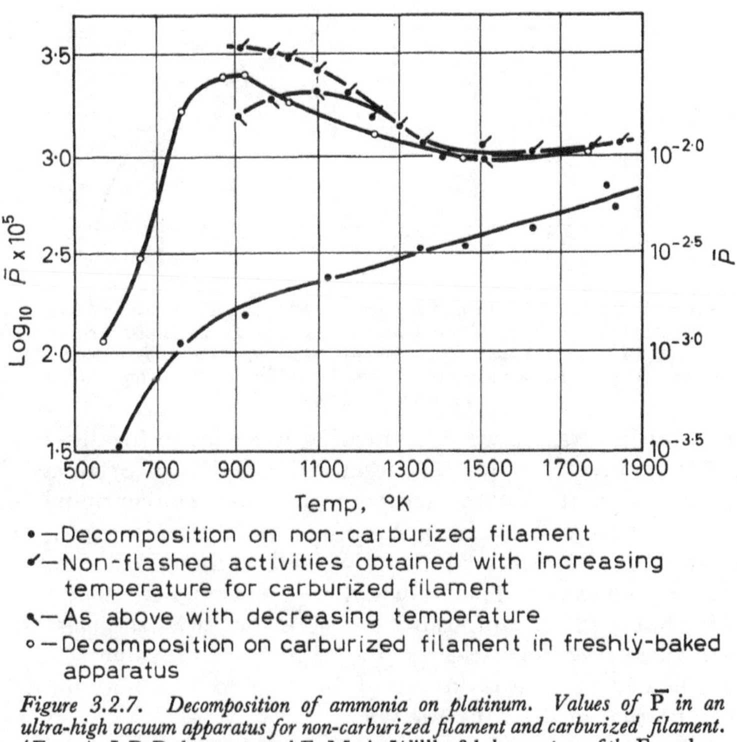

Figure 3.2.7. Decomposition of ammonia on platinum. Values of $\bar{P}$ *in an ultra-high vacuum apparatus for non-carburized filament and carburized filament.* (*From* A. J. B. Robertson and E. M. A. Willhoft[1], *by courtesy of the* Faraday Society)

Clearly the oxygen was helping to produce NO, which was contributing to mass 30. The reaction proceeded with high efficiency and no NO_2 was observed to be present.

That the large negative temperature coefficient of the reaction probability, observed in the $\bar{P}(T)$ plot of hydrazine, was not necessarily a characteristic of N_2H_2 formation, is suggested by the results summarized in *Figures 3.2.7* and *3.2.8*, which show the effect of carburizing a platinum wire on the decomposition of ammonia under comparable conditions. Base pressures in these experiments were less than 2×10^{-9} torr.

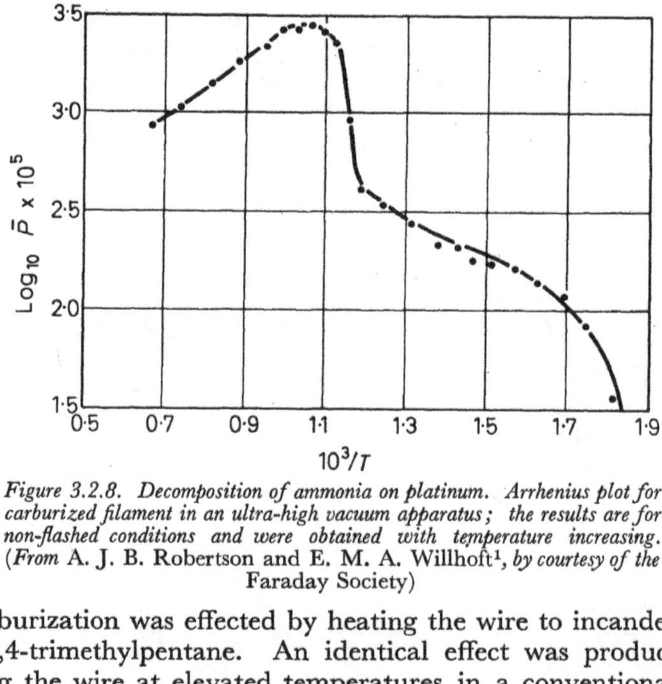

Figure 3.2.8. Decomposition of ammonia on platinum. Arrhenius plot for carburized filament in an ultra-high vacuum apparatus; the results are for non-flashed conditions and were obtained with temperature increasing. (*From* A. J. B. Robertson and E. M. A. Willhoft[1], *by courtesy of the* Faraday Society)

Carburization was effected by heating the wire to incandescence in 2,2,4-trimethylpentane. An identical effect was produced by heating the wire at elevated temperatures in a conventional high vacuum, as was used in the preliminary part of this work. The explanation of the effect of the dissolved carbon in the wire is discussed elsewhere[1]. It is sufficient to mention here that in the case of ammonia decomposition, where no free radicals were observed mass spectrometrically, the significant negative temperature coefficient of $\bar{P}$ was found to be due to small amounts of dissolved carbon in the catalytic wire.

REFERENCES

1 ROBERTSON, A. J. B. and WILLHOFT, E. M. A. *Trans. Faraday Soc.* 63 (1967) 476
2 DIBELER, V. H., FRANKLIN, J. L. and REESE, R. M., *Advances in Mass Spectrometry*, J. D. WALDRON (Ed.), Pergamon, London, 1959
3 FONER, S. N. and HUDSON, R. L. *J. chem. Phys.* 29 (1958) 442
4 FONER, S. N. and HUDSON, R. L. *J. chem. Phys.* 28 (1958) 719

3.3

REARRANGEMENT PROCESSES IN THE FRAGMENTATION OF ORGANIC IONS

J. H. Beynon, A. E. Fontaine and A. E. Williams

3.3.1. INTRODUCTION

MANY compounds exhibit mass spectra containing fragment ions whose origin cannot be described by simple cleavage of bonds in the molecular ion. These ions can only be explained by intra-molecular atomic rearrangement at the moment of fragmentation. The degree of rearrangement may be small, as in the case of a simple hydrogen atom migration or it may be multi-step and involve concerted electron shifts at each stage. In some cases the mass peak caused by 'rearrangement ions' may be very large and may possibly even be the base peak of the mass spectrum.

High resolution, mass spectrometry giving unequivocal empirical formulae of mass measured ions, isotopic labelling at specific sites in the molecules, mainly with deuterium but increasingly often with ^{13}C, ^{15}N and ^{18}O, ionization and appearance potential measurements and the detection of metastable decompositions are all experimental techniques which have been brought to bear on the problem of elucidating the mechanisms of rearrangement-fragmentation ion reactions.

3.3.2. FACTORS INFLUENCING ION ABUNDANCE IN MASS SPECTRA

Newly-formed molecular ions will, according to quasi-equilibrium theory, have a range of excited states, a fraction of which correspond to states having the necessary activation energy for the ion to decompose unimolecularly in a certain way. The rate of each decomposition mode is proportional to the number of activated ions. Similarly, a fraction of the fragment ions of a particular ion configuration may be activated enough to decompose further, again in a variety of ways.

The most abundant fragment ions in a mass spectrum must correspond to ions having the complementary qualifications of high stability and being the result of a 'favourable' decomposition path. The decomposition path to a stable fragment ion or neutral species will be a favourable one, i.e. product stability is a 'driving force' of the decomposition. Other factors can influence the rate of a given fragmentation-rearrangement, the most important being the possibility of an ion attaining a favourable spatial configuration for the reaction to proceed. Other related factors, which are of importance, are the availability of empty orbitals for the intra-ionic formation of new bonds and also the possibility of concerted electron shifts occurring to produce bond fissions.

The relative stabilities of two fragment ions partially determines their relative abundance in the mass spectrum. The various factors influencing ionic stability are, inevitably, also rate determining features of the ion decomposition and rearrangement modes. Thus the participation of a normally non-bonded electron in the formation of a new bond, can give very stable ions, e.g. *m/e* 31, $C_2H=\overset{+}{O}H$, in aliphatic oxygenated compounds and *m/e* 30, $CH_2=\overset{+}{N}H_2$ in aliphatic amines. Ions can sometimes stabilize by resonance in the same way as molecules. Examples of this are to be found in ions which can be considered as having aromatic type structures, e.g. the ion of *m/e* 139 $(C_{11}H_7)^+$, abundant in the spectra of many fused ring hydrocarbons and substituted anthraquinones, possibly has the highly resonant structure:

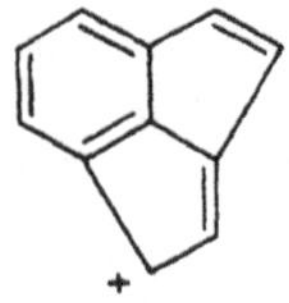

Similarly, enhanced ion stability occurs when a ring structure can be formed from a straight-chain compound. This is especially true of four-, five- and six-membered rings, e.g. *m/e* 135, 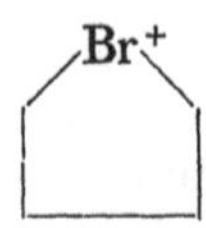 in long-chain alkyl bromides[1]. Even-electron ions having all electrons paired, formed by homolytic cleavage of a bond in the molecular ion, can be expected to be more stable than odd-electron ion radicals. The latter can be formed from the molecular ion only by multiple homolytic cleavage or rearrangement, or both. If the odd-electron radical, produced at the same time as the even-electron ion, is also stable, e.g. NO• or allyl, it is usually found that the decomposition is favoured. Similarly, abundant odd-electron ions are very often produced because the complementary even-electron neutral fragment is a small, highly stable molecule, e.g. H_2O, HCN, CO, C_2H_4, etc. It is also possible that the odd-electron ion may be resonance stabilized.

3.3.3. REARRANGEMENTS

By far the most commonly encountered ion rearrangement is that of hydrogen atom migration. Hydrocarbon spectra contain many examples of such migrations. In a hydrocarbon, there are a larger number of atomic arrangements of similar energy which can be produced for a small amount of ion activation energy. Thus, each hydrogen atom has a very high degree of lability and a number of competing fragmentation reactions can occur giving several rearranged fragments. This phenomenon, resulting in a large variety of rearrangement ions, has been called by McLafferty 'random' rearrangement[2].

Examples of specific hydrogen rearrangements are also fairly common. Thus *m/e* 55 is the base peak in the mass spectra of C_5, C_6 and C_7 cyclic ketones[3].

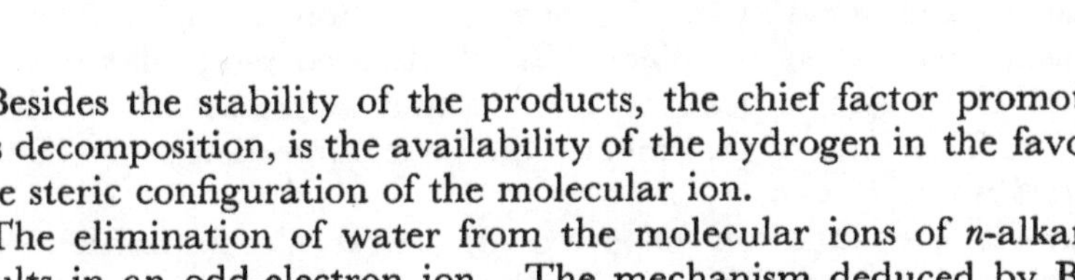

Besides the stability of the products, the chief factor promoting this decomposition, is the availability of the hydrogen in the favourable steric configuration of the molecular ion.

The elimination of water from the molecular ions of *n*-alkanols results in an odd-electron ion. The mechanism deduced by Benz and Biemann[4] and also by Meyerson and Leitch[5] shows that at least 90% of the transferred hydrogen comes from the C_4 position.

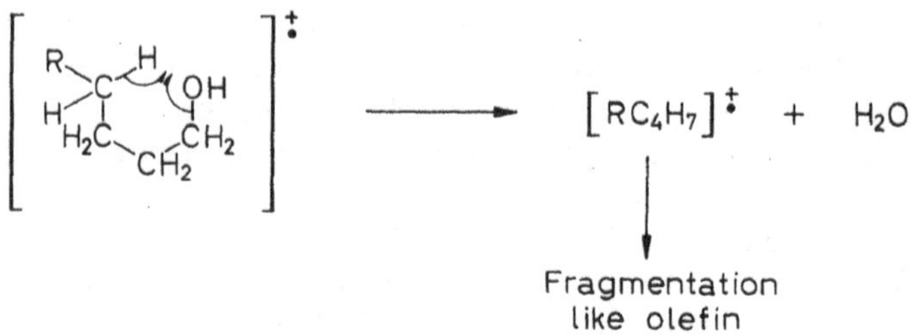

It is possible to formulate a mechanism for this ion decomposition in terms of a favoured six-membered ring transition state. The hydrocarbon fragment ion $[RC_4H_7]^+$ is found to fragment as though it had an olefinic configuration. In such ions a measure of stability is gained by delocalization of the positive charge.

Halogen atoms are known to rearrange though their mobility does not compare with that of hydrogen.

Thus, in the mass spectrum of 1,1,2,3,4,4-hexafluoro-1,4-dibromo-but-2-ene, the ion $[C_3F_3Br_2]^+$ is abundant.

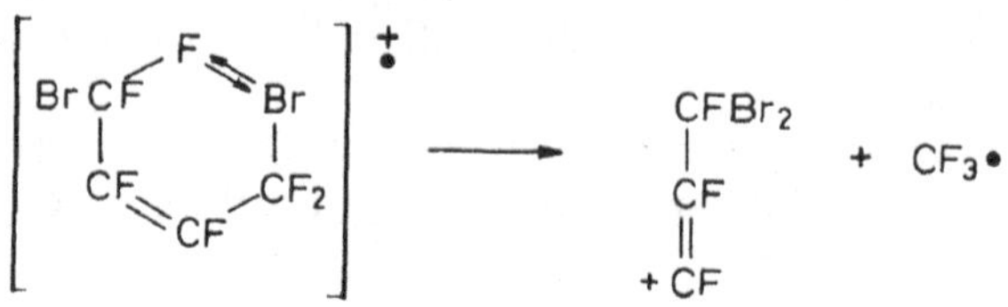

A favourable steric configuration for intra-ionic exchange is possible.

Evidence for unimolecular exchange, prior to fragmentation, is obtained from the mass spectra of aromatic carboxylic acids[7]. Benzoic acid loses ·OH from its molecular ion *m/e* 122 to give a fragment ion at *m/e* 105. The decomposition produces a 'metastable peak' at *m/e* 90·4. Benzoic acid-*d*, $[C_6H_5{\cdot}COOD]$ loses ·OD from its molecular ion *m/e* 123 to give a fragment ion at *m/e* 105 and a 'metastable peak' at *m/e* 89·6. The deuterated molecular ion also loses ·OH to give a peak at *m/e* 106 and a 'metastable peak' at *m/e* 91·3. The ratio of the 'metastable peak' abundances *m/e* 89·6 to *m/e* 91·3 is one to two.

This is the situation expected if, before decomposition, the deuterium could equilibrate with the two hydrogen atoms *ortho* to the rotating carboxylic acid group on the benzene ring. If the positive charge on the molecular ion of benzoic acid is located on the oxygen atoms, it is possible to write an intermediate species such as:

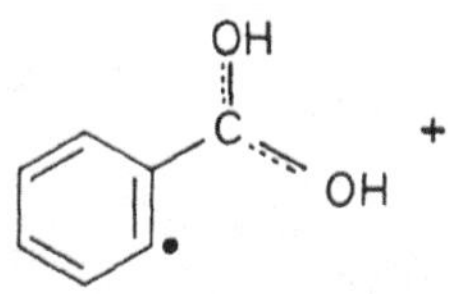

with the charge resonating between the two oxygens.

If the original hydroxyl group is deuterated and the deuterium is able to exchange freely before decomposition via such an intermediate, then the probability of loss of ·OD or ·OH, that can be obtained by comparing the metastable peak abundances, is one to two (*Figure 3.3.1*).

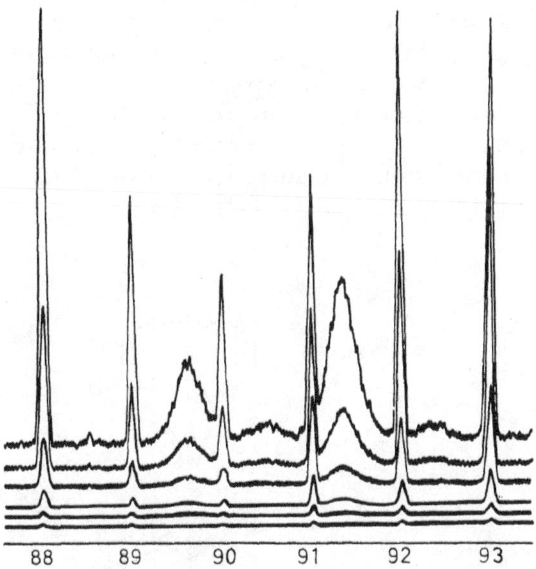

Figure 3.3.1. Metastable peaks in the mass spectrum of deuterobenzoic acid due to loss of ·OD *and* ·OH *from the molecular ion.* (*From* J. H. Beynon, B. E. Job and A. E. Williams[7], *by courtesy of* Zeitschrift für Naturforschung)

Several instances of methyl group rearrangement are known, the earliest reported being by Florin *et al.*[8] in the trimer of dimethylphosphinoborine. The base peak of this mass spectrum is at m/e 41.

The ion responsible is unambiguously identified as $B(CH_3)_2^+$ from the boron isotope pattern.

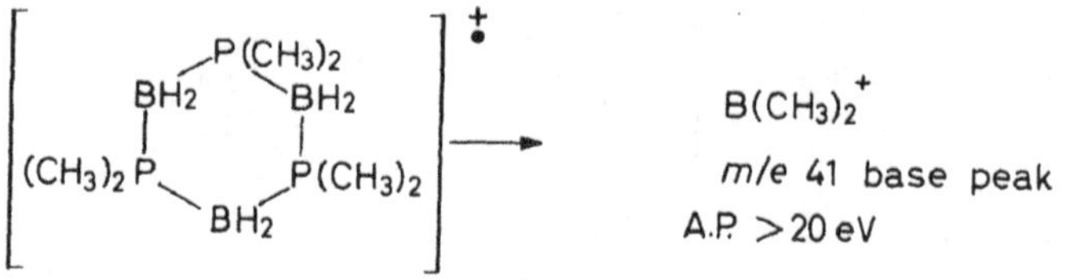

From J. H. Beynon, B. E. Job and A. E. Williams[11]. (*By courtesy of* Zeitschrift für Naturforschung)

It is worthy of notice that the formation of this ion involves the migration of two methyl groups and that the appearance potential of the ion is more than 20 eV, indicating that a considerable amount of activation energy is necessary for the rearrangement to occur.

Another more recent example of methyl migration has been discussed by Djerassi[9, 10]. This example illustrates how fairly subtle structural changes in the molecule can have a spectacular effect on the probability of a decomposition pathway.

One of the abundant peaks in the spectrum of *trans*-Δ^3-10-methyl-2-octalone is at *m/e* 69 and is due to the ion $C_4H_5O^+$.

Evidence obtained from the mass spectra of various deuterated *trans*-Δ^3-10-methyl-2-octalones (II, III, IV):

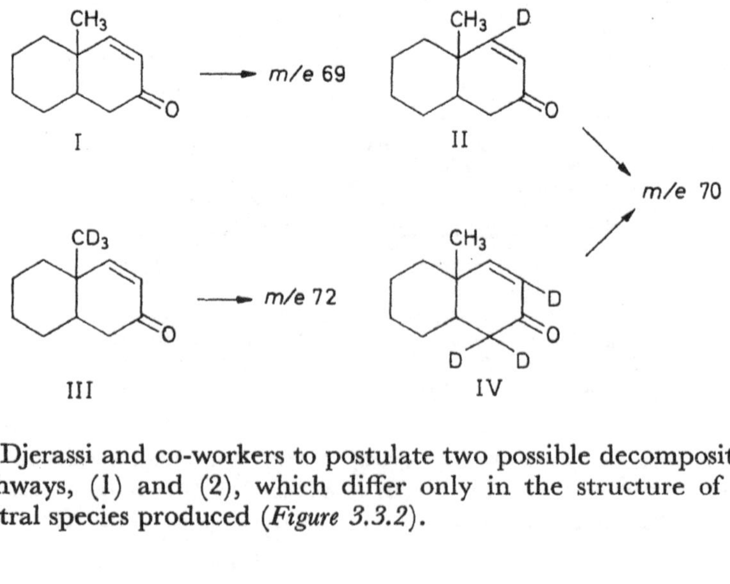

led Djerassi and co-workers to postulate two possible decomposition pathways, (1) and (2), which differ only in the structure of the neutral species produced (*Figure 3.3.2*).

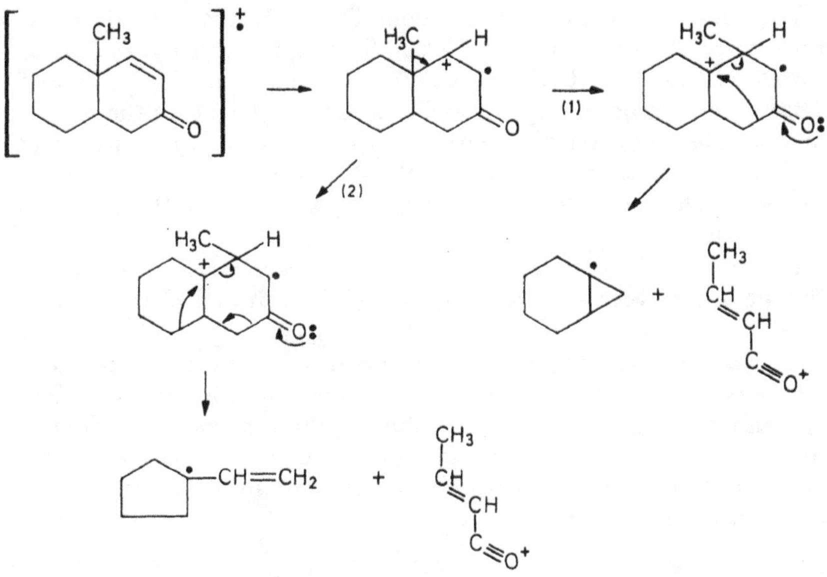

Figure 3.3.2

Methyl migration was observed to only a small extent in the mass spectra of *A* and *B* (*Figure 3.3.3*) but was responsible for a very abundant ion (*m/e* 69) in the mass spectra of *C* and *D*.

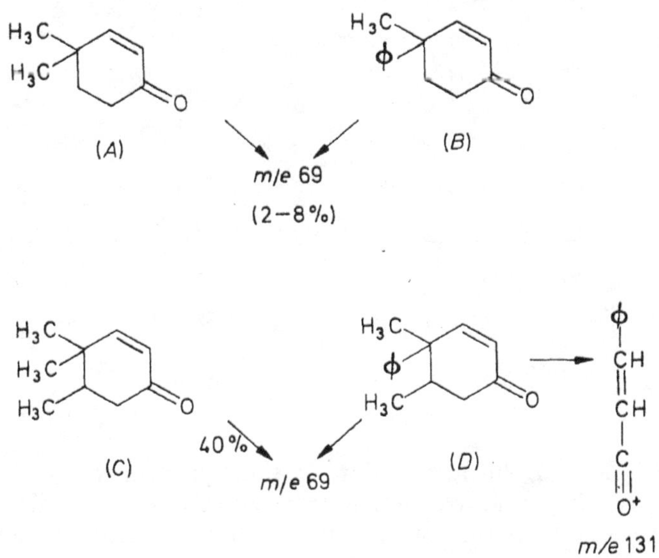

Figure 3.3.3

The mass spectrum of *D* also contains an ion of *m/e* 131, many times more abundant than *m/e* 69, indicating that phenyl migration is preferred to methyl. Thus the effect of substitution at position 5 of the cyclohexenone ring, produces a marked effect on the ease of the methyl–phenyl rearrangement process. This fact indicates that (2) is the acceptable rationale for the decomposition of *trans*-Δ^3-10-methyl-2-octalone, since it involves the participation of the substituent at position 5.

The influence of structure on a decomposition pathway is also evident in the work of Beynon *et al.*[11] on substituted nitronaphthalenes.

A prominent ion in the mass spectrum of 1-nitronaphthalene is that caused by the loss of carbon monoxide from the molecular ion. As 2-nitronaphthalene and 1,8-dinitronaphthalene do not lose carbon monoxide from their molecular ions, the indication is that the 8-position is concerned in the decomposition routes. The proposed mechanism is:

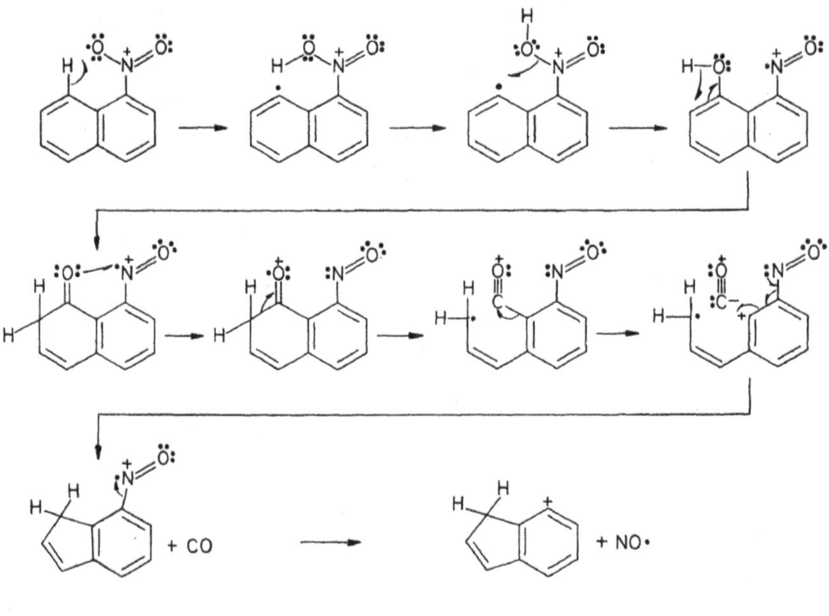

and depends on the abstraction of the hydrogen atom from the 8-position. The ease with which this first step occurs is dictated by the nature and position of substituents on the naphthalene system. The electron density at the 8-position, calculated according to where a nitro group(s) is positioned, is, in order of decreasing charge density:

1,3-dinitro > 1-nitro > 1,4-dinitro > 1,5-dinitro

In 1,3-dinitronaphthalene the electron density at the 8-position is so high that the OH-migration back to the 8-position, as required by the proposed mechanism, is partially suppressed and this compound shows a very abundant ion corresponding to the loss of •OH as well as some tendency to lose CO from the molecular ion. Otherwise the elimination of the CO• fragment occurs with a probability in accordance with the series. For 1,5-dinitronaphthalene there is zero loss of CO. The mass spectra of some nitronaphthalenes are given in *Figure 3.3.4.*

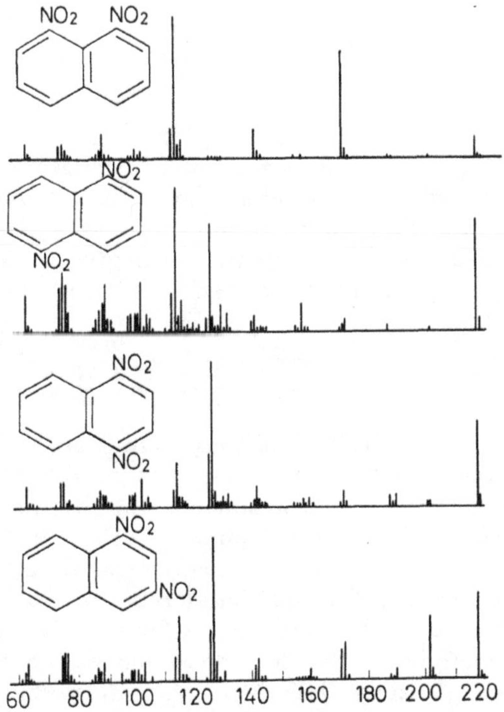

Figure 3.3.4. *The mass spectra of some nitronaphthalenes.* (*From* J. H. Beynon, B. E. Job and A. E. Williams[11], *by courtesy of* Zeitschrift für Naturforschung)

A very intriguing rearrangement is that involving the loss of neutral nitric oxide from an ion. A large number of organic nitro compounds eliminate NO• from their molecular ions or from an ion formed by the elimination of another small neutral fragment[12], or both.

The driving force is doubtless the great stability of the NO· radical and the resonant even-electron ion produced. All the 21 compounds studied by Beynon *et al.* also give large peaks at *m/e* 30. These are caused by the very stable nitric oxide positive ion.

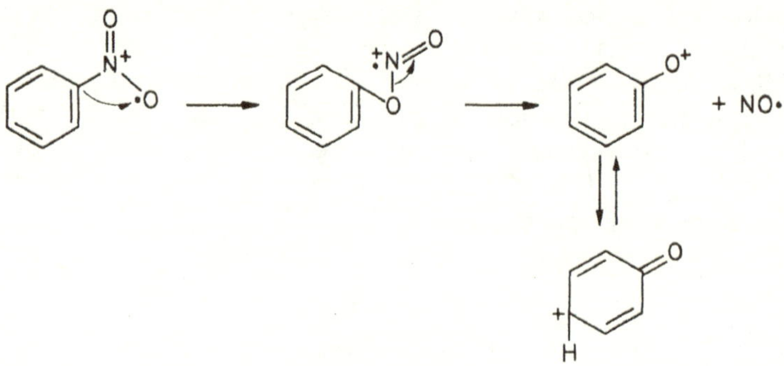

Figure 3.3.5

Figure 3.3.5 shows a possible formulation for the loss of NO· from nitrobenzene. Isomerization of the nitro group to the nitrite form was put forward by Beynon *et al.* as a possible first step leading to the loss of NO· from the nitro group. The loss of NO^+ can be explained in a similar way.

3.3.4. STUDY OF REARRANGEMENTS BY METASTABLE PEAKS

The decomposition of ions having an average lifetime of approximately 10^{-6} s (metastable) can be detected in mass spectrometry since, in most instruments, the decomposition products cause 'metastable peaks' to appear in the mass spectrum[13].

Metastable peaks are easily recognized in a mass spectrum since they are broader and considerably weaker than normal mass peaks. The broadening is almost certainly caused by the release of a range of energies in the decompositions of the metastable ions (0 eV to approximately 0·05 eV). Almost nothing is known about the distribution(s) of the released energies which occur, but the fact that almost all rearrangement-fragmentations are accompanied by metastable peaks shows that their average life times are much longer than simple single bond, fission fragmentations.

Occasionally a rearrangement-fragmentation occurs with the release of a much larger amount of energy (0·2–2 eV). The metastable peaks observed for these fragmentations can have widths of several mass numbers and have flat or dished tops[14, 15].

Wide metastable peaks are also known to occur for fragmentations which are apparently simple single bond fissions, e.g. in dimethyl guanidine:

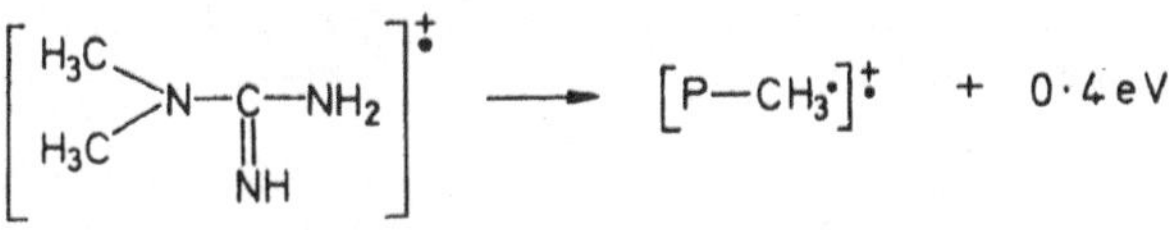

The reason why some ion rearrangement-fragmentations occur with the release of relatively large energies is thought to be due to the extremely high stability of both fragments of the decompositions. Thus in *ortho-*, *meta-* and *para*-nitrophenols the loss of NO· from the molecular ions of the *ortho* and *para* forms is accompanied by 0·75 eV of energy[14]. For the *meta* form a large release of energy is not detected. Stable resonant structures can be written for the fragment ions of the *ortho* and *para* cases but not for the *meta* (*Figure 3.3.6*).

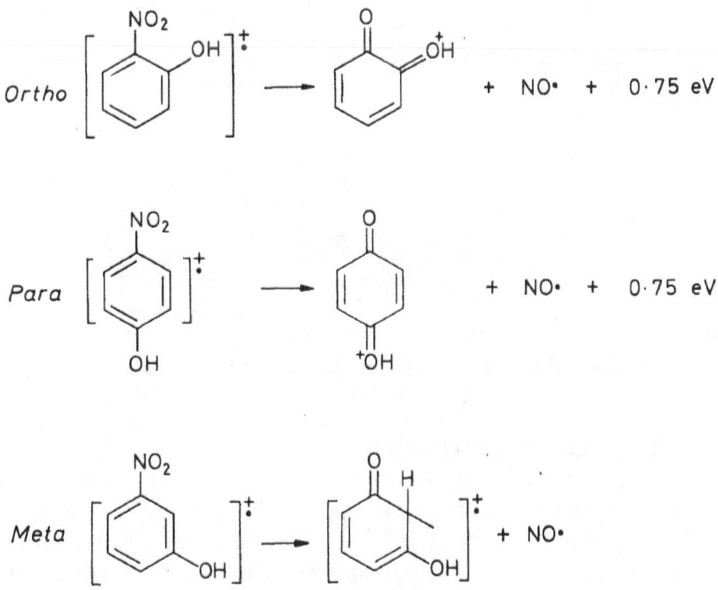

Figure 3.3.6. Suggested structures for the ions caused by loss of NO· *from the molecular ions of the nitrophenols*

The consecutive loss of neutral carbon monoxide from the molecular ion of anthraquinone to yield the molecular ions of fluorenone and biphenylene, was studied and reported some years ago by Beynon *et al.*[16]. This rearrangement has, as its driving force, the high stability of the CO neutral fragment and the resonance stabilized fluorenone and biphenylene ions (*Figure 3.3.7*). These ions dominate the anthraquinone spectrum and have been studied at low ionizing electron energies. The rearranged ions were found to occur in abundance, very close to the ionization potential of the anthraquinone molecule.

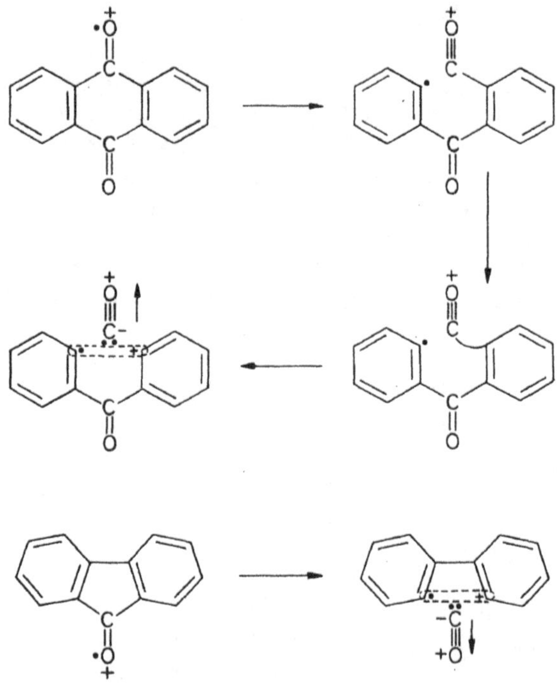

Figure 3.3.7. The consecutive loss of CO *from anthraquinone*

It is clear that the activation energies required for some rearrangements are very small indeed, but there are some, such as already mentioned in the mass spectrum of the dimethylphosphinoborine trimer, that require much more activation. There must be some correlation between activation energies of ion rearrangements and the relative importance of the various stability and steric factors which influence the decomposition path.

The type of rationale used by the mass spectroscopist, in attempting to explain the intra-ionic bonding changes which must occur in ion rearrangements, is very similar to that used by the organic chemist in discussing inter- and intra-molecular rearrangements. There are similarities between unimolecular molecular ion rearrangement dissociations observed in mass spectra, and other types of unimolecular decomposition, those most often quoted being pyrolysis and photolysis. The possibility of ion rearrangements indicating possible preparative paths to the chemist has not been overlooked. Brown *et al.*[17] have already described the similarity between the mass spectral and pyrolytic fragmentations of some carbonyl compounds and have given an efficient route to biphenylene.

REFERENCES

1 McLafferty, F. W. *Analyt. Chem.* 34 (1962) 2
2 McLafferty, F. W. *Appl. Spectrosc.* 11 (1957) 148
3 Beynon, J. H., Saunders, R. A. and Williams, A. E. *Appl. Spectrosc.* 14 (1960) 95
4 Benz, W. and Biemann, K. *J. Am. chem. Soc.* 86 (1964) 2375
5 Meyerson, S. and Leitch, L. C. *J. Am. chem. Soc.* 86 (1964) 2555
6 McLafferty, F. W. *Analyt. Chem.* 31 (1959) 82
7 Beynon, J. H., Job, B. E. and Williams, A. E. *Z. Naturf.* 20A (1965) 883
8 Florin, R. E., Wall, L. A., Mohler, F. L. and Quinn, E. I. *J. Am. chem. Soc.* 76 (1954) 3344
9 Kamitsky Jr., F., Gurst, J. E. and Djerassi, C. *J. Am. chem. Soc.* 87 (1965) 1398
10 Djerassi, C., *Some Newer Physical Methods in Structural Chemistry,* United Trades Press, London, 1967
11 Beynon, J. H., Job, B. E. and Williams, A. E. *Z. Naturf.* 21A (1966) 210
12 Beynon, J. H., Saunders, R. A. and Williams, A. E. *Industrie chim. belge* 4 (1964) 311
13 Hipple, J. A. and Condon, E. U. *Phys. Rev.* 69 (1946) 347
14 Beynon, J. H., Saunders, R. A. and Williams, A. E. *Z. Naturf.* 20A (1965) 180
15 Beynon, J. H. and Fontaine, A. E. *Z. Naturf.* 22A (1967) 334
16 Beynon, J. H. and Williams, A. E. *Appl. Spectrosc.* 14 (1960) 156
17 Brown, R. F. C., Gardner, D. V., McOmnie, J. F. W. and Solly, R. K. *Chem. Commun.* 8 (1966) 407

3.4

A NOVEL ION IN THE MASS SPECTRA OF ARYLUREAS AND RELATED COMPOUNDS

A. G. Loudon, A. MacColl, V. Seddon, D. Smith, K. S. Webb and A. Ribera

RECENTLY the formation of six-membered ring ions under electron impact has been recognized and reported[1, 2]. Apart from our preliminary communication[3] only one possible formation of a five-membered ring ion has been reported[2], although there are other compounds whose reported fragmentation under electron impact could be explained in this manner[4]. In this paper we wish to present the evidence for the formation of such five-membered ring ions in a large number of compounds, of general formula, shown in *Table 3.4.1.*

Table 3.4.1. The Types of Compound Studied {$C_6H_4(AB)—X—C(=Y)—Z$}

Compounds	X	Y	Z
Ureas	NH	O	NR_2
	NH	O	NHAr
Thioureas	NH	S	NR_2
	NH	S	NHAr
	NH	S	NH_2
	NH	S	NHAc
Carbamates	NH	O	OCH_3
	O	O	$N(CH_3)_2$
Guanidines	NH	NH	NHAr
	NH	NAr	NHAr

When A is an *ortho* substituent then, in most cases, large $(M—A)^+$ ions can be seen (*Tables 3.4.2.–3.4.5*), but when A is a *meta* or *para* substituent, these ions are very much smaller (less than 0·5%) (*Figure 3.4.1*).

Table 3.4.2. $(M-Cl)^+$ *from* o-$ClC_6H_3(R)NHCONR'_2$. (*After* M. A. Baldwin *et al.*[3], *by courtesy of* Chemical Communications)

R	R′	$(M-Cl)^+$%	$(M-Cl)^+/(M)^+$
H	Me	36	3·4*
H	*i*-Pr	8	0·6*
4-Me	Me	38	2·7*
4-Me	Et	33	1·4*
5-Cl	Me	24	5·6*
5-Cl	Et	25	3·7*
5-Cl	*n*-Pr	19	2·6*
6-Me	Me	62	9·2*

* These values have been corrected for the isotopic contribution at (M + 2).

Table 3.4.3. $(M-Cl)^+$ *from* o-$ClC_6H_3(R)NHCONH\ (C_6H_3R'R'')$. (*After* M. A. Baldwin *et al*[3], *by courtesy of* Chemical Communications)

R	R′	R″	$(M-Cl)^+$%	$(M-Cl)^+/(M)^+$
H	H	H	8·5	0·4*
6-Me	H	H	32·0	1·5*
H	2-Me	H	5·3	0·2*
H	2-Cl	H	7·9	0·7*
H	2-MeO	H	3·8	0·07*
3-Cl	2-Cl	3-Cl	3·6	0·8*
5-Cl	2-Cl	5-Cl	11·0	1·8*
6-Me	2-Cl	6-Me	50·0	12·8*

* These values have been corrected for the isotopic contribution at (M + 2).

Table 3.4.4. $(M-A)^+$ *from* o-AC_6H_4 NHCSNHR′

R′	A	$(M-A)^+$%	$(M-A)^+/(M)^+$
H	F	4·0	10
Ac	F	18·0	36
H	Cl	100·0*	> 100
Ac	Cl	31·0*	> 100
H	Br	100·0*	> 100
Ac	Br	11·0*	> 100
H	I	†*	—
Ac	I	16·0*	> 100
Ac	NO_2	33·0*	> 100

* No M+.
† M—HI ⪢ M—I.

Only when Y = S is there any large loss of hydrogen, presumably from an *ortho* position. The *ortho*halogenophenyl-N′-acetyl thioureas

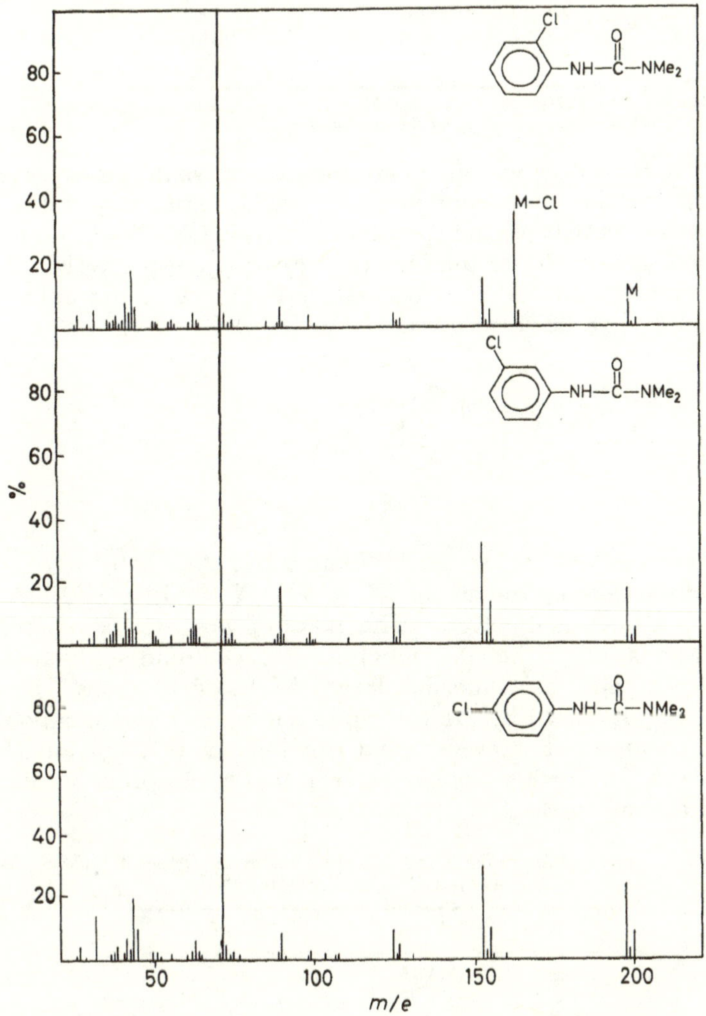

Figure 3.4.1. Mass spectra of o, m *and* p *substituted chlorophenyl ureas*

and the corresponding thioureas show also (M−HX)$^+$ ions (X = Br, I) which in the latter case, overwhelms the loss of iodine. In the N′-aryl-N′-acetyl thioureas the M−X and M−HX ions both show a subsequent loss of ketene.

Table 3.4.5. (M−Cl)$^+$ *from o*-$ClC_6H_3(R)OCONMe_2$. (*After* M. A. Baldwin *et al.*[3], *by courtesy of* Chemical Communications)

R	(M−Cl)$^+$%	(M−Cl)$^+$/(M)$^+$
3-Cl	7·4	18·8*
5-Cl	8·4	12·0*
6-Cl	5·3	3·6*

The compound *o*-$ClC_6H_4NHCOOCH_3$ has (M−Cl)+ = 100% and (M−Cl)+/(M)+ = 2·2.
* These values have been corrected for the isotopic contribution at (M + 2).

If the *ortho* effect was an electronic one, it would be expected to operate for the *para* compounds as well. The only reasonable explanation is that of ring formation, in which X, Y or Z could be involved, giving rise to the ions (*a*), (*b*) or (*c*) respectively. As all the monohaloanilines show a small (M−A)$^+$ ion, it would seem that ion (*c*) can be ruled out (X = NH).

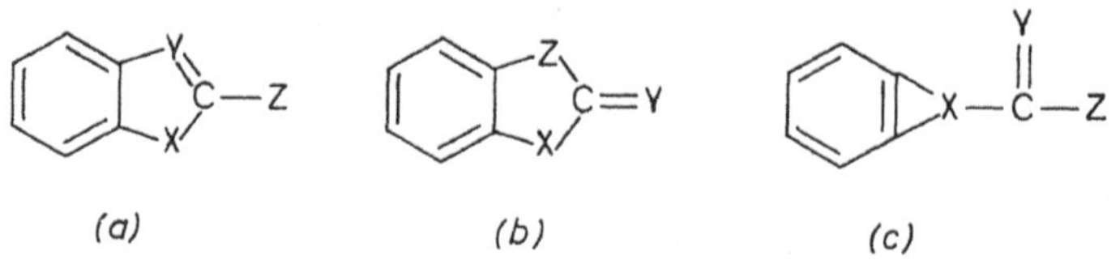

On the basis of the work of Williams and Waight[1, 2] it was first considered that in the ureas (X = NH, Y = O, Z = NHAr or NR_2), ion (*a*) was involved. This point of view was reinforced by the examination of the *ortho*monohalogenoacetanilides which showed, with exception of fluorine, large (M–halogen)$^+$ ions. Similar ions were not shown by the corresponding *meta* or *para* compounds. It was pointed out, however, that this did not rule out ion (*b*) in other cases and further work has shown that this is probably the ion sometimes involved.

Table 3.4.6. Effect of Increasing Size of Alkyl Groups on the (M−Cl)$^+$/(M)$^+$ *Ratio in Compounds o*-$ClC_6H_3(R')NHCONR_2$

R	R′=H	R′=Cl
Me	3·4*	5·6*
Et	1·5*	3·7*
n-Pr	—	2·6*
i-Pr	0·6*	—

*These values have been corrected for the isotopic contribution (M + 2).

This was investigated by increasing the size of the alkyl groups in the N-aryl-N′-dialkyl compounds of general formula *o*-Cl·$_6CH_3$-R′·NH·CO·N·R_2 (R= Me, etc., see *Table 3.4.6*). This will produce both steric and electronic effects and the complications of such an approach are revealed by examination of the unimolecular rate theory as applied to mass spectrometry. McLafferty[5] has derived the following equation:

$$\frac{(\mathrm{M} - \mathrm{A})^+}{(\mathrm{M})^+} = k_1/(\Sigma k_1' + \Sigma k_{\mathrm{inst.}}) \qquad \ldots (3.4.1)$$

where k_1 is the rate constant for the formation of the $(M-A)^+$ ion from the molecular ion (M^+) and $\Sigma k_1'$ is the sum of all the rate constants for the subsequent decomposition of the $(M-A)^+$ ion. $\Sigma k_{inst.}$ reflects a rather ill-defined term for any instrumental discrimination between various ions. In this case $\Sigma k_{inst.}$ can be regarded as effectively constant since the focusing properties of the machine should not change very much when only slightly different ions are being considered. However, both k_1 and $\Sigma k_1'$ will be affected by electronic and steric modifications, which complicate the interpretation of the results.

The effect of increasing the size of the alkyl groups could be a steric one if ion (*b*) was involved, since interaction between the alkyl groups and the benzene ring would lessen the stability of ion (*b*). Electronically, the greater stabilization of the $(M-A)^+$ ion, due to the increasing electron donating powers of the bigger groups, would be, in part at least, opposed by their stabilizing effect on any ions produced by subsequent fragmentation of this ion.

Comparison with the thioureas helps to elucidate the structure involved. In the simple *o*-monochlorophenyl thiourea the $(M-Cl)^+$ ion is the base peak but the molecular ion is so small that the $(M-Cl)^+/(M)^+$ ratio cannot be measured accurately, so a direct comparison is impossible. However, the phenyl-N′-dialkylthioureas show fairly large $(M-H)^+$ ions, presumably formed by the same mechanism. The large increase in the $(M-H)^+$ ion in the thioureas, compared to the ureas, suggests the same mechanism is operating.

Table 3.4.7. Effect of Increasing Size of Alkyl Groups on $(M-H)^+/(M)^+$ *in Compounds* $C_6H_5NHCSNR_1R_2$

$R_1=R_2$		$R_1=H$	
R_1	$(M-H)^+/(M)^+$	R_2	$(M-H)^+/(M)^+$
H	11	H	11
Me	13	Me	9
Et	16	Et	9
n-Pr	18·5	*n*-Pr	11
n-Bu	20	—	—

In contrast to the ureas mentioned above, where the $(M-Cl)^+/(M)^+$ ratio goes down with increasing size of the alkyl group (*Table 3.4.6*), the $(M-H)^+/(M)^+$ ratio goes up with increasing size of the alkyl groups (*Table 3.4.7*). With the monoalkyl compounds it is effectively constant.

The four possible structures for the ions in these two cases are a_1, a_2 and b_1, b_2:

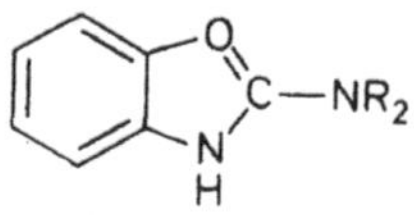

(a_1)

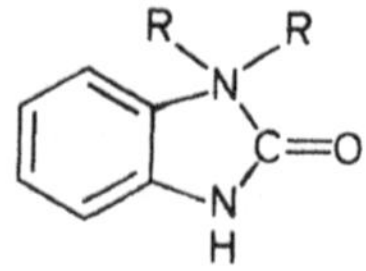

(b_1)

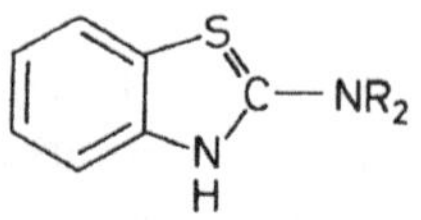

(a_2)

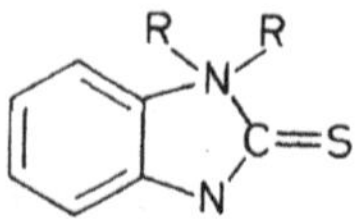

(b_2)

The quite noticeable change in the magnitude of the ratio $(M-Cl)^+/(M)^+$ in the corresponding ureas and thioureas, suggests that ions (b_1) and (b_2) cannot both be involved. The difference in behaviour on increasing the size of the alkyl groups, suggests that both (a_1) and (a_2), or, indeed, both (b_1) and (b_2) cannot be involved. This leaves the two possible combinations (a_1) and (b_2), or (a_2) and (b_1). The relative stabilization of such ions might be expected to depend on the polarizability of the lone pair of electrons which interacts with the electron deficient aryl nucleus. It is difficult to reconcile this with the pair (a_1) and (b_2), since the lone pair of sulphur is more easily polarized than that of oxygen. That is to say, if (a_1) is correct, then (a_2) would be expected, but this combination has been ruled out. Thus, by default, the two ions must be (a_2) and (b_1).

The $(M-H)^+/(M)^+$ ratios for some thioureas and corresponding N′-acetyl thioureas are shown in *Table 3.4.8*. The fact that these are very similar for a considerable change in the electronic environment of the far nitrogen atom is strong evidence for the correctness of this assignment. It would be a surprising coincidence if the alterations in k_1 and $\Sigma k_1'$, induced by the acetyl group, were such as to produce *in toto* such a small change. This would be required if the structure (b_2) were correct.

Table 3.4.8. Ratio $(M-H)^+/(M)^+$ *for* (A)$C_6H_4NHCSNHR_1$ *Showing Similarity in the Behaviour of the Thioureas and the Corresponding Acetates*

A	$(M-H)^+/(M)^+$	
	R_1=H	R_1=Ac
o-F	0·3	0·1
m-F	0·4	0·3
m-Br	0·2	0·1
m-Cl	1·5	1·2

Another effect, which seems purely steric in nature, is shown in the N-aryl and NN′-diaryl ureas. The presence of methyl substituents in the 6-position increases the $(M-A)^+/(A)^+$ ratio enormously. This may be due to the bulk of the methyl group forcing the side chain to curl round more often to the other side of the benzene nucleus and hence speeding up the production of the $(M-A)^+$ ion.

In this case, where any changes in the molecule ion other than in the nature of A, are reflected in the $(M-A)^+$ ion, McLafferty's approach[5] cannot be used to investigate substituent effects, since both k_1 and $\Sigma k_1'$ vary simultaneously. This double variation may explain the fact that $(M-Cl)^+/(M)^+$, in the case of 2,5-dichloroacetanilide, is higher than for 2-chloracetanilide but which is higher than for 2-chloro-5-nitroacetanilide, whereas superficially the two groups would be expected to weaken, in the carbon–chlorine bond, by electron withdrawal. In general, the effect of a chlorine *para* to the leaving chlorine seems to be to increase the $(M-Cl^+)/(M)^+$ ratio (*Tables 3.4.2* and *3.4.3*), which is also the case in 2,5-dichlorophenyl-N′-dimethylcarbamate.

If, however, the leaving group is varied to give the same ion, equation 3.4.1 can be simply modified to give equation 3.4.2, since in this case $\Sigma k_1'$ and $\Sigma k_{inst.}$ must be constant:

$$\ln [(M-A)^+/(M)^+] = \ln B - \Delta E/RT + C \quad \ldots (3.4.2)$$

If it is assumed that B, the frequency factor, is a constant, equation 3.4.2 can be used to see if the rate-determining step in this reaction is the fission of the carbon–A bond. If this is the case the plot of log $[(M-A)^+/(M)^+]$ against the appropriate bond energy should be a straight line. A series of simple *o*-substituted acetanilides were examined. The graph shown in *Figure 3.4.2* shows that this is not the case here, but there is a reasonably straight line for the halogens. The discrepancy in the other cases could be due to the rate-determining step not being the sp^2 bond fission, or to

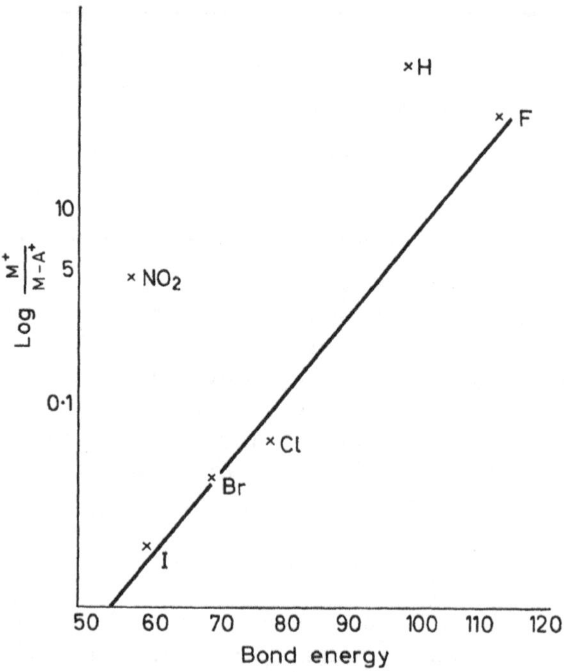

Figure 3.4.2. Plot of log (M − A)$^+$/M$^+$ *against bond energy for a series of substituted acetanilides*

the frequency factor *B* not being constant, or to the fact that, in the nitro compound, the bond energy used is that determined for nitromethane and there may be a considerable difference between the sp^2 and sp^3 bond energies. The last seems unlikely, since it would require a difference of 40 kcal between the two bond energies. Thus, although some points have been elucidated, there is a considerable amount of work still to do on the urea, thiourea and acetanilide compounds. In particular it is hoped to study the appearance potentials of these ions.

The guanidines have been less fully explored. The *ortho*monochlorophenyl compound shows a small (M—Cl)$^+$ ion which is not shown by the *meta* or *para* compound. Monophenylguanidine and the monochlorophenylguanidines also show (M—H)$^+$ ions ($<$ 10%).

The structure of the monophenyl molecular ion is not clear, it could be either (*d*) or (*e*), leading to the $(M-A)^+$ ion (*f*), (*g*) or (*h*).

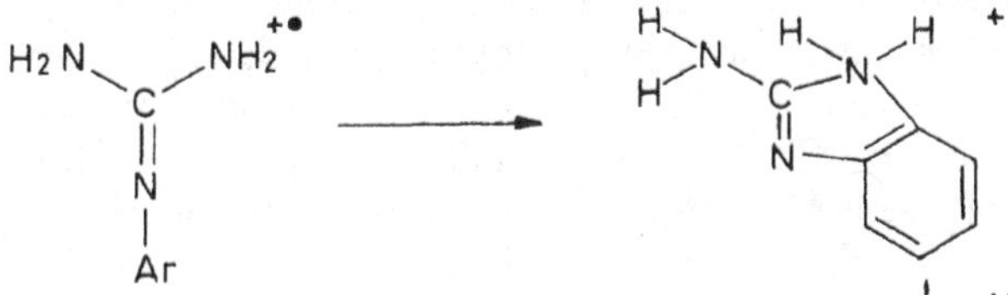

(*d*) (*f*)

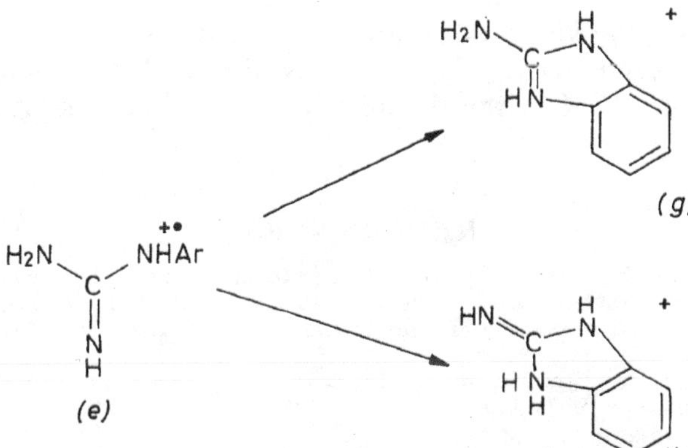

(*e*) (*g*) (*h*)

It is interesting to note that in the case of symmetrical triphenylguanidine, the $(M-H)^+/(M)^+$ ratio rises to 25%; in this case the ion (*i*) is probably involved. This difference suggests that in the monophenyl compounds ion (*f*) is involved. Finally, the loss of a chlorine atom, in the case of some *o*-chlorophenyl-N′-dimethylsulphonamides studied, is much smaller. The reason for this is not clear.

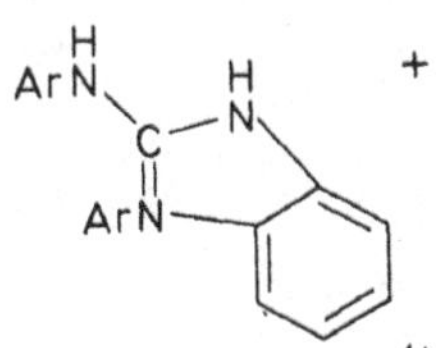

(i)

Before concluding, one or two other fragmentations of these compounds are worth noting. In many of the non-alkylated thiourea compounds and the corresponding N′-acyl compounds there is an ion corresponding to the appropriate thiophenol. This is akin to the formation of phenylmethylsulphide from phenylmethylthiocarbamate[6]. There is no evidence of alkyl migration to give sulphides in the N-aryl-N′-dialkyl thioureas. The observed formation of $(M-45)^+$ ions in the *N*-arylmethylcarbamates is in accordance with the work of Lewis on ethyl compounds[7], but has not yet been reported for the parent compound, probably as a result of pyrolysis[6].

It seems that this *ortho* effect probably covers a wide range of compounds and is less dependent on which hetero-atom is present, than is the case in some rearrangement ions[6]. The detailed mechanisms have yet to be completely elucidated and may be quite different in each case. The problems discussed show some of the difficulties involved in the detailed interpretation of fragmentation patterns.

REFERENCES

1 WILLIAMS, D. H., *Newer Physical Methods in the Study of Structural Chemistry*, Laboratory Practice, p. 11, London, 1967 also WAIGHT, E. S. *ibid.* p. 67

2 RONAYNE, J., WILLIAMS, D. H. and BOWIE, J. A. *J. Am. chem. Soc.* 88 (1966) 4980

3 BALDWIN, M. A., LOUDON, A. G., MACCOLL, A., SMITH, D. and RIBERA, A. *Chem. Commun.* (1967) 350

4 JÖNG, J., HOURIET, R. and SPITELLER, G. *Mh. Chem.* 97 (1966) 1004

5 BURSEY, M. M. and MCLAFFERTY, F. W. *J. Am. chem. Soc.* 88 (1966) 529

6 THOMSON, J. B., BROWN, P. and DJERASSI, C. *J. Am. chem. Soc.* 88 (1966) 4049

7 LEWIS, C. P. *Analyt. Chem.* 36 (1964) 176 and 1582

3.5

MASS SPECTRA OF CYCLOTETRAZENOBORANES

E. F. H. Brittain, J. B. Leach and J. H. Morris

THE mass spectra of two series of cyclotetrazenoboranes have been studied. The first series are substituted in the 2,5 positions and the second series have an additional substituent on the boron atom. The structures are:

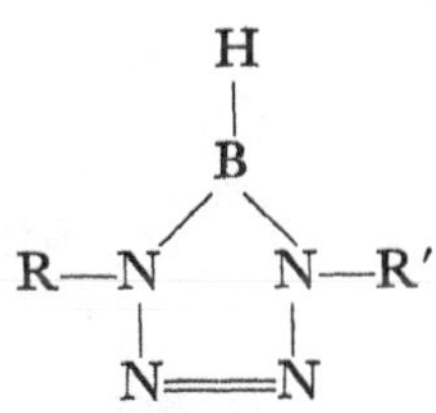

Series 1

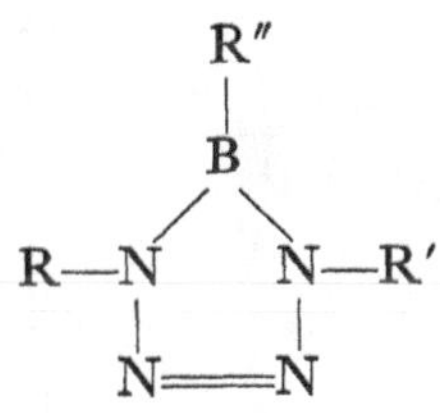

Series 2

Series 1 contained R = R′ = Me; R = R′ = Et; R = Me, R′ = Ph and R = R′ = Ph. Series 2 consisted of R = R′ = R″ = Et; R = R′ = Ph, R″ = Me and R = R′ = R″ = Ph.

The interpretation of the mass spectral breakdown of the molecules is facilitated by the isotopic abundance of boron since ^{10}B and ^{11}B occur in the ratio 19·6:80·4%. This gives a triplet in the parent and fragment ion peaks due to contributions from what we have called the parent peak (*P*) ($^{10}B^{12}C^{14}N$), the $P + 1$ peak ($^{11}B^{12}C^{14}N$, $^{10}B^{13}C^{14}N$) and the $P + 2$ peak ($^{11}B^{13}C^{14}N$). In calculating peak intensity ratios the ^{15}N isotope has also been included. The major peak is always the $P + 1$ or $F + 1$ peak (F = fragment).

Details of the breakdown have been confirmed in most cases, by metastable transitions, where it is normally possible to observe two metastable peaks, one due to the transition involving the ^{10}B and the other involving the ^{11}B isotopes.

3.5.1. DIMETHYLCYCLOTETRAZENOBORANE

Figure 3.5.1 shows the fragmentation pattern of 2,5-dimethylcyclotetrazenoborane. The parent peaks at 97, **98** and 99* are in accordance with the expected relative intensity ratios. The peak at **98** is

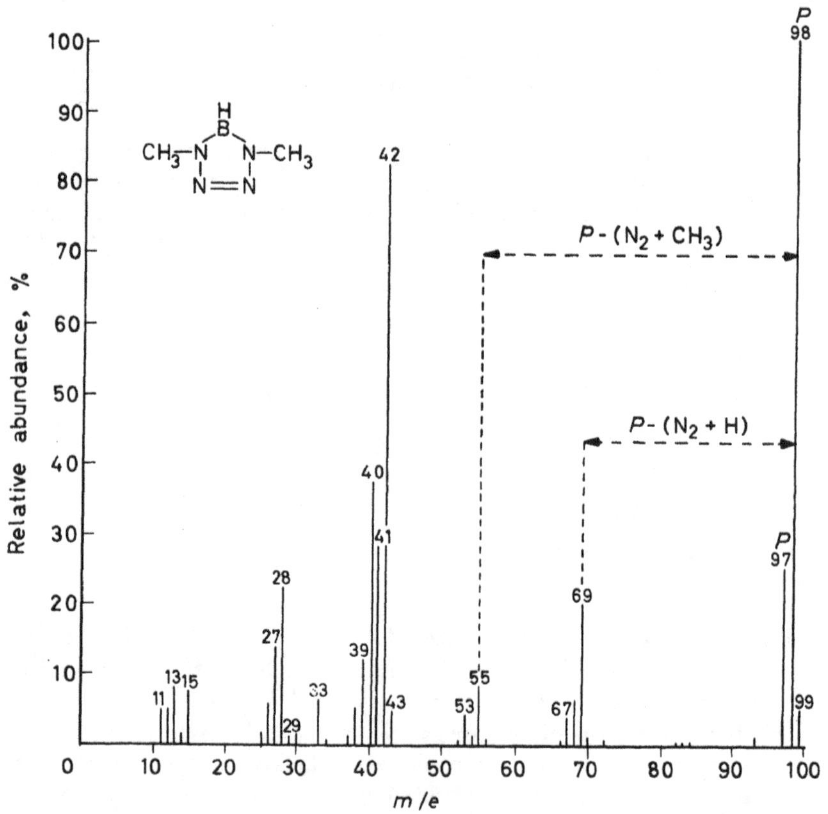

Figure 3.5.1. Fragmentation pattern of 2,5,-dimethylcyclotetrazenoborane. (*From* E. F. H. Brittain *et al. by courtesy of the* Chemical Society)

the base peak in the spectrum. Earlier work by Morris and Perkins[1] suggests that the initial ionization is by loss of an electron from one of the lone pairs of the azo-nitrogen atoms.

---

* The major peak, due to ^{11}B is in bold type throughout this paper.

The loss of $N_2 + H^{\bullet}$ is confirmed by metastables to give, we believe, the species:

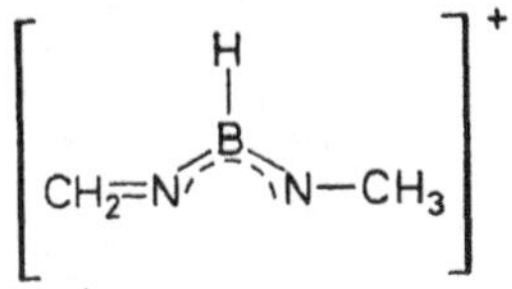

m/e 68, **69**

I*a*

or the cyclic alternative:

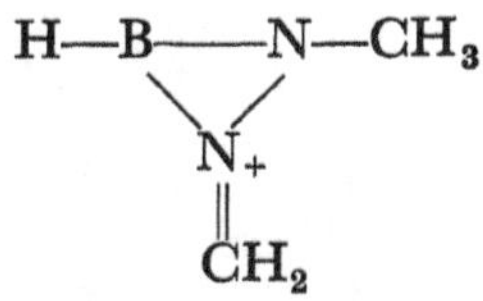

I*b*

A very small peak at *m/e* 83 indicates that loss of $CH_3^{\bullet}$ from the molecular ion can also occur.

The major fragmentation of I occurs by loss of HCN, with accompanying transfer of $H^{\bullet}$, to give the prominent peak at *m/e* 42 (II).

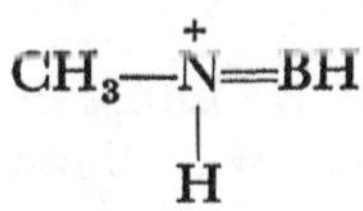

m/e 41, **42**

II

The peak at *m/e* 55 is thought to be due to loss of ($N_2 + CH_3^{\bullet}$) although this is not confirmed by a metastable transition. The species formed probably has the structure III, since loss of HCN then occurs, confirmed by the appropriate metastable transitions.

$$[H—N—BH—N{=}CH_2]^+$$

m/e 54, **55**

III

3.5.2. DIETHYLCYCLOTETRAZENOBORANE

Figure 3.5.2 shows the fragmentation pattern of 2,5-diethylcyclotetrazenoborane. Again the parent peak due to ^{11}B is the base peak

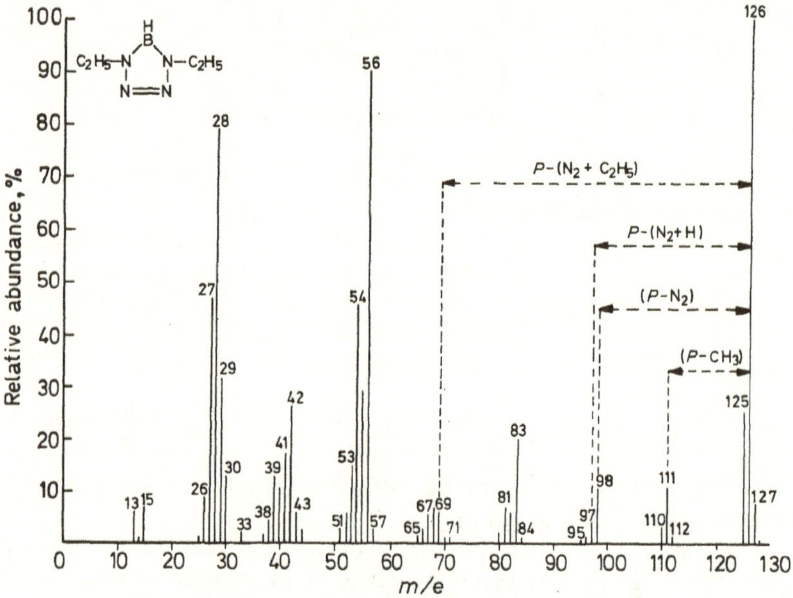

Figure 3.5.2. Fragmentation pattern of 2,5-diethylcyclotetrazenoborane. (*From* E. F. H. Brittain *et al. by courtesy of the* Chemical Society)

of the spectrum, and the relative intensities of the peaks at *m/e* 125, **126** and 127 are in agreement with those expected on the basis of isotopic distribution.

$CH_3^{\bullet}$ is lost fairly readily from the ethyl side-chain to produce a peak 110, **111** and 112. The resulting even-electron species is:

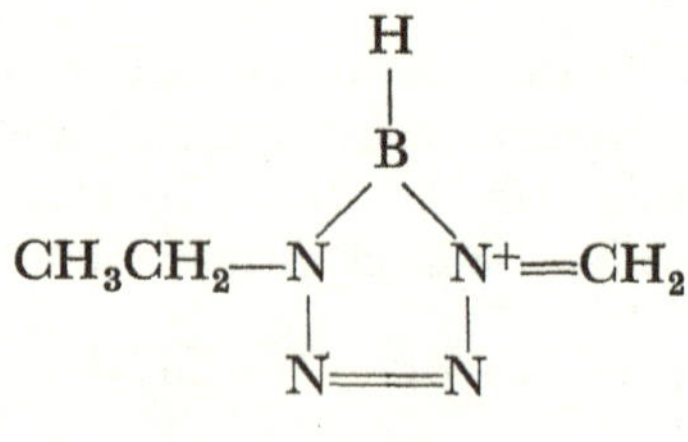

m/e 110, **111,** 112

IV

Loss of N_2 also occurs from the parent to give:

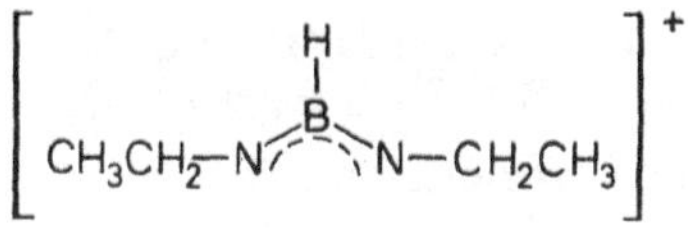

m/e 97, **98**, 99

V

However, the intensity ratios for 97, **98** and 99 show a high value for 97 so we suggest that, as in 2,5-dimethylcyclotetrazenoborane, there is also a loss of (N_2 + H•). In most of the compounds studied, N_2 is lost first and then there is a metastable peak corresponding to the loss of H.

Loss of $N_2 + C_2H_5^{\bullet}$ is confirmed by metastable peaks to produce peaks at 68, **69** and 70.

Peaks at *m/e* 82, **83** and 84 are produced by $CH_3^{\bullet}$ loss from V. This is then followed by loss of HCN to contribute to the peaks at *m/e* 55, **56** and 57. To account for this fragmentation, we suggest that, when N_2 is lost from the parent ion, transfer of a hydrogen atom can occur to give species VII as an alternative configuration to V.

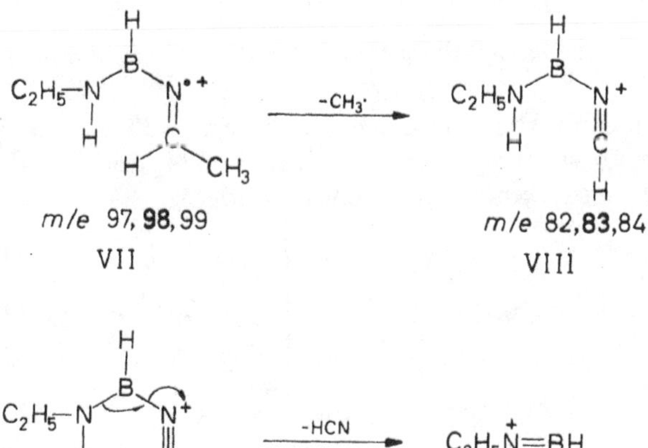

m/e 97, **98**, 99 — VII; *m/e* 82, **83**, 84 — VIII

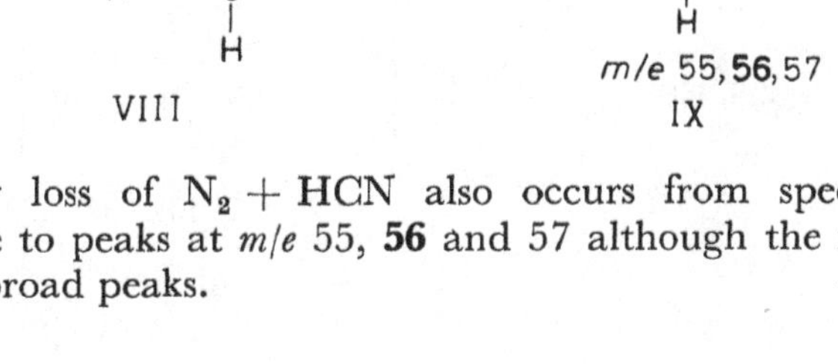

VIII; *m/e* 55, **56**, 57 — IX

Possibly loss of N_2 + HCN also occurs from species IV to contribute to peaks at *m/e* 55, **56** and 57 although the metastables are very broad peaks.

There are thus two possible routes to *m/e* **56**, which probably accounts for its very high abundance.

3.5.3. 2-METHYL-5-PHENYLCYCLOTETRAZENOBORANE

Figure 3.5.3 shows the fragmentation pattern. The parent peak at *m/e* 160 is again very strong and the associated peaks at *m/e* 159 and 161 are in the correct ratio.

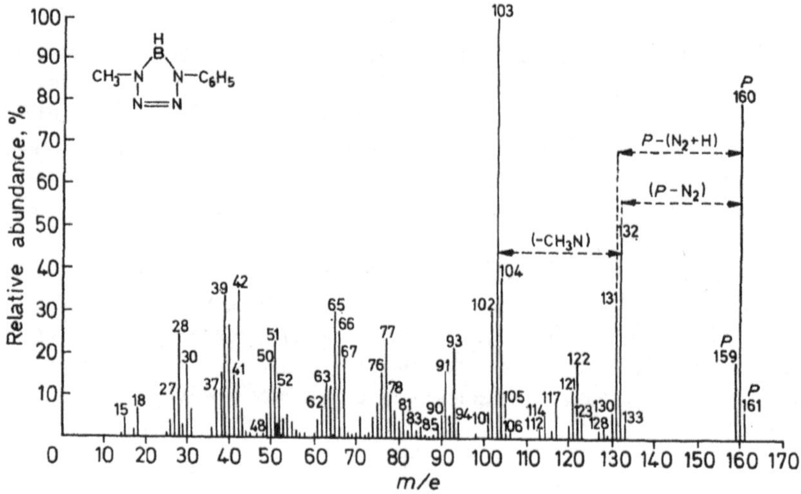

Figure 3.5.3. Fragmentation pattern of 2-methyl-5-phenylcyclotetrazenoborane. (*From* E. F. H. Brittain *et al. by courtesy of the* Chemical Society)

It is noticeable that the peaks at 131 and 132 are comparable in abundance, so we suggest that both loss of N_2 and $N_2 + H\bullet$ occur, as in the dimethyl and diethyl compounds.

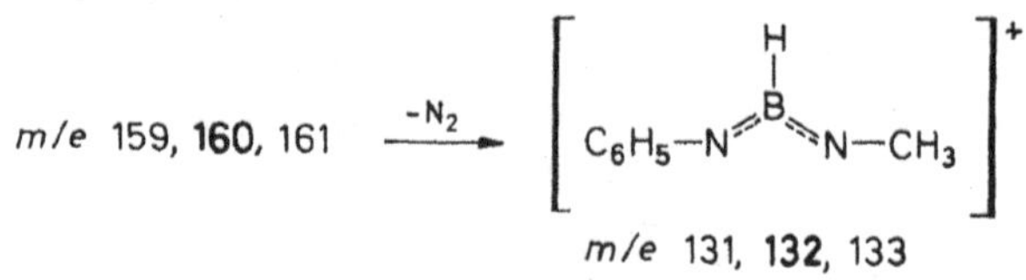

X

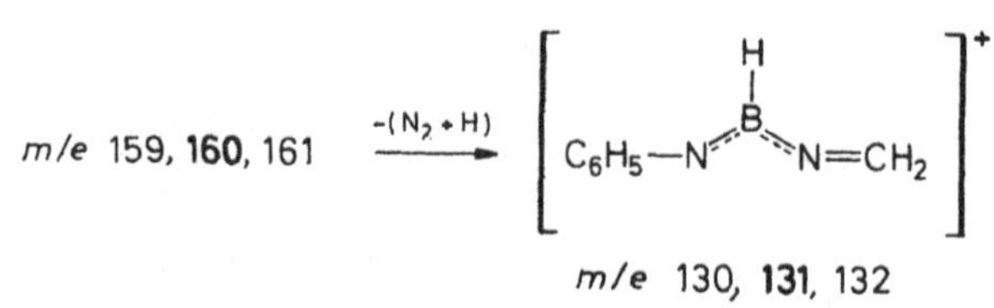

XI

Loss of CH_3N from X produces the base peak at *m/e* 103.

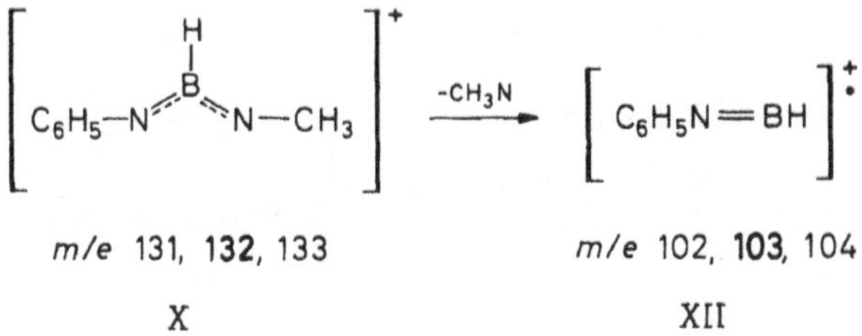

The peak at *m/e* 93 is probably due to aniline produced by thermal decomposition in the heated inlet system.

3.5.4. 2,5-DIPHENYLCYCLOTETRAZENOBORANE

The fragmentation pattern is shown in *Figure 3.5.4.* Parent ion peaks at *m/e* 221, **222** and 223 are still quite intense and are in the expected ratios.

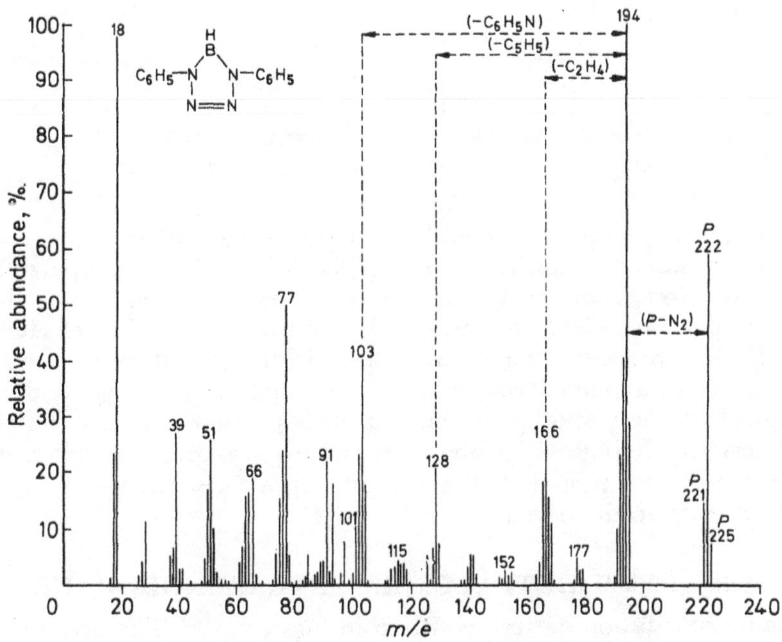

Figure 3.5.4. Fragmentation pattern of 2,5-diphenylcyclotetrazenoborane. (*From* E. F. H. Brittain *et al. by courtesy of the* Chemical Society)

Fragmentation of the molecular ion now occurs by loss of N_2 predominantly:

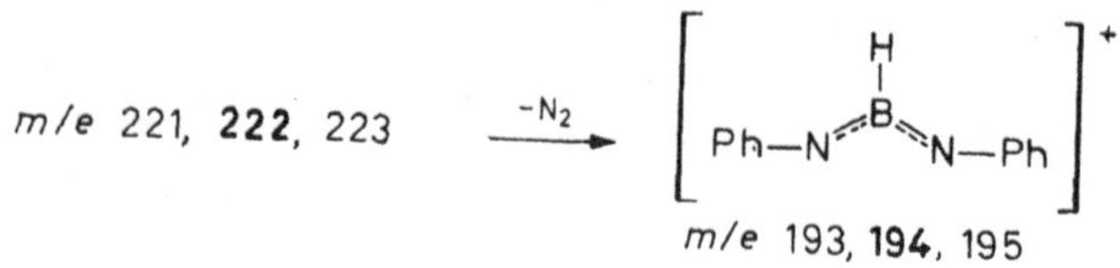

XIII

but, since *m/e* 193 has an abundance of about 42% rather than 25%, loss of $N_2 + H\cdot$ can also occur.

Many of the lower *m/e* peaks can be accounted for in terms of the breakdown of the phenyl rings, particularly if the rings open on electron impact as has been suggested by Reed. Fragmentation at successive carbon atoms in the phenyl system accounts for peaks centred on 177, 166, 154, 141, 128 and 117. Most of the cleavages are supported by metastable transitions.

Loss of PhN from XIII could give peaks at *m/e* 102, **103** and 104:

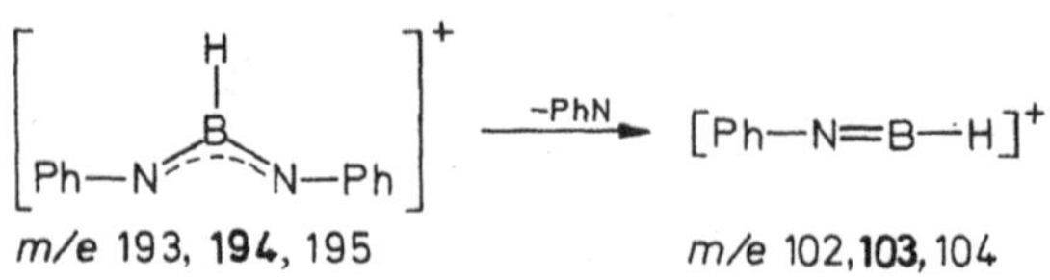

XIII XIV

although this is not confirmed by metastable transitions.

We consider that during the loss of N_2 and H in these compounds, H is not lost from the boron but from the side chain, since the subsequent breakdown patterns and, in particular, the loss of HCN fit this interpretation somewhat better. In the case of the diphenyl, H loss from a phenyl ring could occur, and *m/e* 193 would then fragment by loss of other sections of the Ph group as does *m/e* 194. To test this hypothesis deuteration studies are being carried out. We have also prepared the second series of compounds with substituents on the boron.

3.5.5. 1,2,5-TRIETHYLCYCLOTETRAZENOBORANE

The fragmentation pattern is shown in *Figure 3.5.5.* The base peak is again the parent ion containing ^{11}B, as it was in the 2,5-diethyl compound.

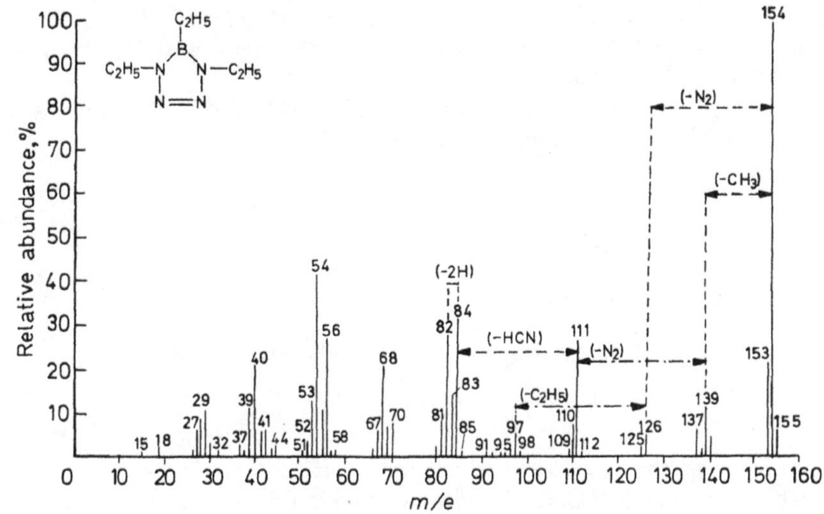

Figure 3.5.5. Fragmentation pattern of 1,2,5-triethylcyclotetrazenoborane

Again methyl loss occurs to give peaks at *m/e* 138, **139** and 140 for which we suggest the structure:

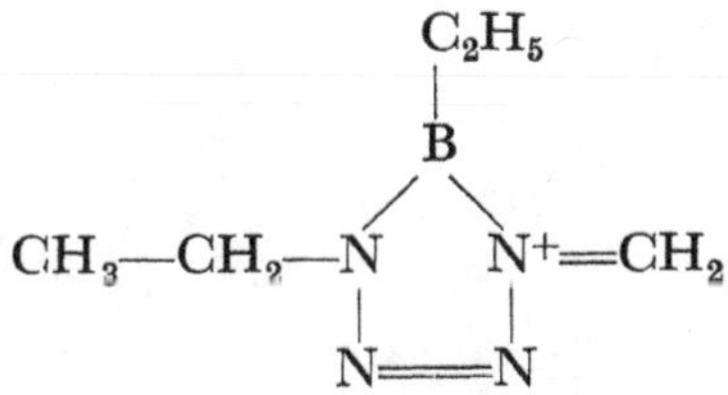

m/e 138, **139,** 140

XV

Loss of N_2 occurs from the parent ions to give peaks at 125, **126** and 127 although, due to the size of the 125 peak, loss of $N_2 + H\cdot$ probably occurs as well, even though ethyl is now on the boron.

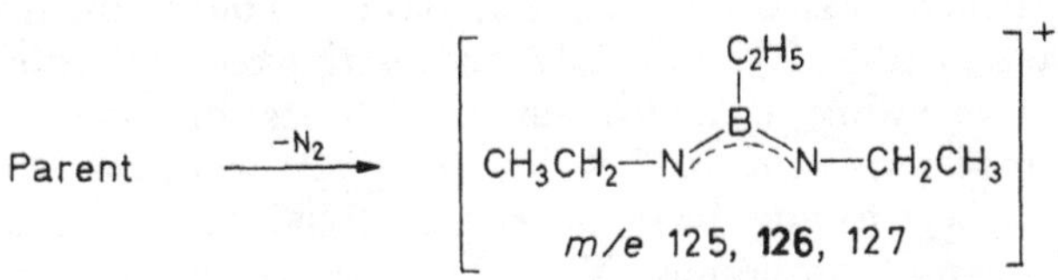

XVI

XVI will lose ethyl to produce peaks at *m/e* 96, **97** and 98, although this does not now occur simultaneously with N_2 loss, as in the diethyl compound.

XV will lose N_2 to give peaks at 110, **111** and 112 which will then lose HCN to give *m/e* 83, **84** and 85.

m/e 138, **139**, 140 — XV   *m/e* 110, **111**, 112 — XVII

If, during the N_2 loss, a hydrogen atom is transferred, then it is fairly easy to lose HCN. Species XVII would become:

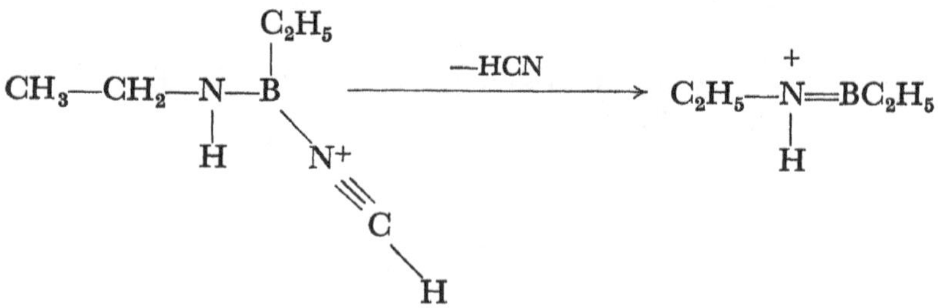

m/e 83, **84**, 85

XVIII

All of which is comparable with the diethyl compound, except that there is no metastable in the triethyl compound, corresponding to the simultaneous loss of N_2 and HCN.

3.5.6. 1-METHYL-2,5-DIPHENYLCYCLOTETRAZENOBORANE

The mass spectrum is given in *Figure 3.5.6.* The intensities of the parent peaks at 235, **236** and 237 are in the correct ratios. It is extremely interesting that loss of $CH_3^{\bullet}$ does not occur, which supports our statement that H is not easily lost from the boron.

N_2 is again lost to give peaks at *m/e* 207, **208** and 209 and again, as in the diphenyl compound, $N_2 + H^{\bullet}$ loss occurs, since the peak at 207 is too intense to arise solely from N_2 loss.

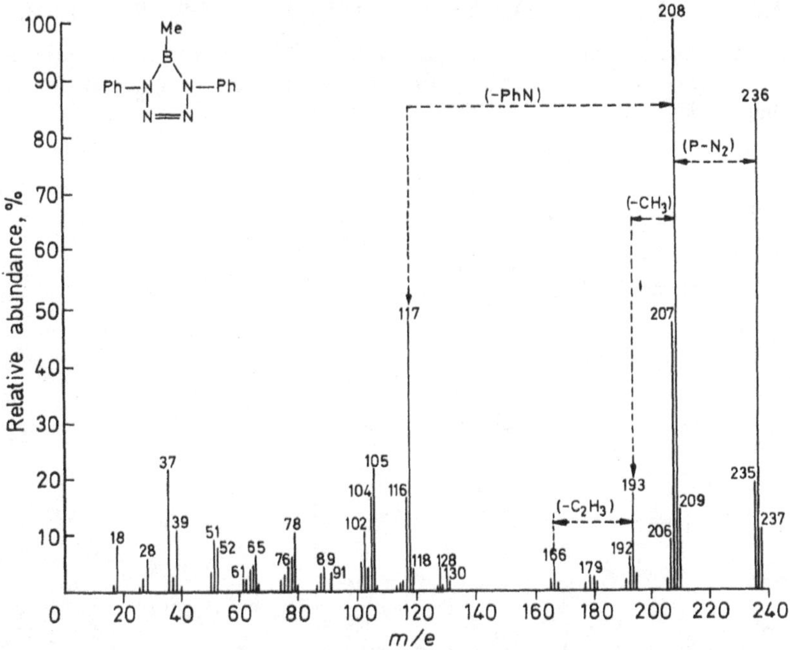

Figure 3.5.6. Fragmentation pattern of 1-methyl-2,5-diphenylcyclotetrazenoborane

The other fragmentation of major interest is loss of PhN from the peaks at 207, **208** and 209 to give the species:

$$[\text{Ph—N=B—Me}]^{+\cdot}$$

m/e 116, **117,** 118

XIX

although this is not confirmed by metastable ions.

In this case, however, loss of Me from *m/e* 207 and **208** is confirmed by metastable peaks, to give species at 192 and **193**. The only reasonable postulate is that Me is now lost from the boron to give:

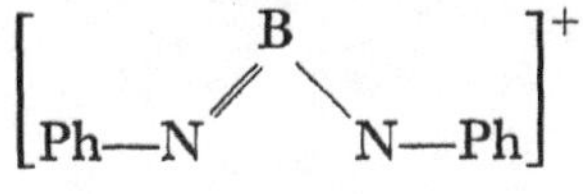

m/e 192, **193,** 194

XX

although it does not seem likely that a positive charge will reside on the boron.

3.5.7. 1,2,5-TRIPHENYLCYCLOTETRAZENOBORANE

The spectrum is shown in *Figure 3.5.7.* The parent ion peaks at *m/e* 297, **298** and 299 are a little less intense than in the diphenyl compound.

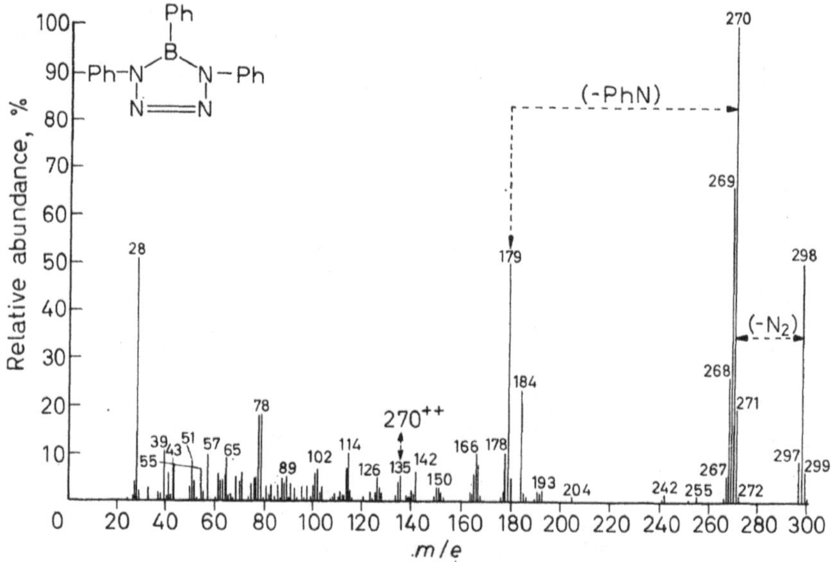

Figure 3.5.7. Fragmentation pattern of 1,2,5-triphenylcyclotetrazenoborane

Loss of N_2 occurs to give peaks at 269, **270** and 271 and, since the *m/e* 269 peak is more intense than expected, loss of $N_2 + H\cdot$ occurs as well. This again supports our contention that H is not lost from the boron in the diphenyl compound.

The other significant loss is of PhN, from the $P - N_2$ species. This loss is not confirmed by metastable peaks and gives rise to the 178, **179** and 180 triplet. The peak at *m/e* 179 has been mass measured and has proved to be $C_{12}H_{10}N^{11}B$ as expected. The structure is presumably:

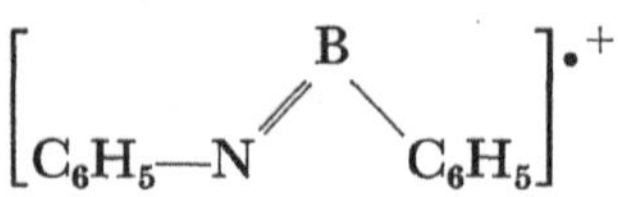

m/e 178, **179,** 180

XXI

In a private communication, Paetzold of Munich has confirmed that ArNBAr is stable as a monomer.

The peak at *m/e* 184 is thought to be a possible impurity peak since its intensity was very much reduced when a second sample was prepared.

3.5.8. 1,2,5-TRIMETHYLCYCLOTETRAZENOBORANE

In the last few days we have managed to prepare 1,2,5-trimethylcyclotetrazenoborane. Its spectrum is not shown since it contains an impurity peak at *m/e* 93. The base peak, however, is the ^{11}B parent ion at *m/e* 112 and the peaks at *m/e* 111 and 113 are in the correct ratio.

The most significant feature is a large peak (20%) due to loss of $N_2 + H^\bullet$ with no peak corresponding to loss of N_2.

3.5.9. SUMMARY

We feel justified in claiming that when loss of $N_2 + H^\bullet$ occurs, the H loss is from the side chain and not from the boron and that the B—R bond is remarkably stable under electron impact conditions.

The loss of N_2 or $N_2 + H^\bullet$ also confirms that the compounds are substituted in the 2,5 position and not in the 3,4 position which is the other possible alternative.

3.5.10. EXPERIMENTAL

Mass spectra of Me_2N_4BH, Et_2N_4BH, $Me(Ph)N_4BH$, and Me_2N_4BMe were obtained on an AEI MS2 mass spectrometer, operating at 70 eV, using a gallium inlet system at 120°C. Source pressures were in the range $0{\cdot}4$–$1{\cdot}6 \times 10^{-6}$ mm.

The spectrum of Ph_2N_4BH was obtained using a direct insertion probe and improved metastable results for the compounds examined on the MS2 were obtained by using an AEI MS12.

The spectra of Et_2N_4BEt, Ph_2N_4BPh and Ph_2N_4BMe were obtained on an AEI MS9 instrument, using a direct insertion probe.

Materials were prepared by the methods described in the literature[2, 3].

Metastable peaks were allocated using a programme on an Elliott 4120 computer.

3.5.11. ADDENDUM

We have now prepared the deuterated derivative:

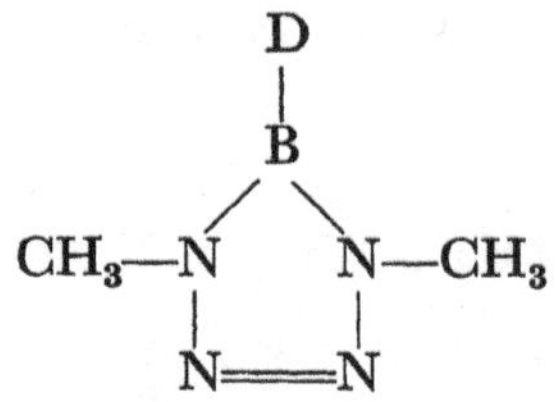

The mass spectrum still shows a loss of 29 u which confirms our interpretation that the initial loss of $N_2 + H^\bullet$ from the dimethyl compound, and presumably the others, involves H loss from the side-chain and not from the boron.

The other major fragmentations proposed in the dimethyl compound are confirmed, since in the deuterated compound the corresponding peaks occur 1 u higher.

We gratefully acknowledge the assistance of Dr. Reuben of the University of Surrey, who obtained the spectrum of diphenylcyclotetrazenoborane, and Mr. Paisley of the National Physical Laboratory for his help in obtaining spectra.

REFERENCES

[1] MORRIS, J. H. and PERKINS, P. G. *J. chem. Soc.* (A) (1966) 580
[2] GREENWOOD, N. N. and MORRIS, J. H. *J. chem. Soc.* (1965) 6205
[3] LEACH, J. B. and MORRIS, J. H. In press

3.6

THE DECOMPOSITION OF 9,10-DIPHENYLANTHRACENE UNDER ELECTRON IMPACT

K. R. Jennings and A. F. Whiting

3.6.1. INTRODUCTION

UNSUBSTITUTED aromatic hydrocarbons give mass spectra consisting of intense peaks due to the molecular ions, together with peaks of much lower intensities due to fragment ions. Peaks due to doubly-charged ions are frequently moderately intense and account for a considerable proportion of the total ionization[1], and low-intensity peaks due to triply-charged ions can also be observed[2]. In discussing the fragmentation processes which give rise to the observed 70 eV mass spectra of these substances, therefore, the production of singly-charged fragment ions from doubly- and triply-charged precursors must be considered. Metastable transitions in which two singly-charged ions are formed from a doubly-charged ion, have been observed in the mass spectra of many aromatic hydrocarbons[3] and a recent study of the mass spectrum of benzene suggests that several ions arise primarily from the decomposition of doubly-charged species[4].

In this paper we wish to describe a study of the decomposition of 9,10-diphenylanthracene under electron impact in which fragmentation reactions of singly-, doubly- and triply-charged ions have been investigated.

3.6.2. EXPERIMENTAL

All observations were made using an AEI MS9 double focusing mass spectrometer of Nier–Johnson geometry, which had been modified to allow independent variation of the accelerating voltage V and the electric sector voltage E[5]. This modification enables one to observe metastable transitions at high sensitivity, arising from ions which dissociate in the field-free region between the source and the electric sector. Usually E is kept constant so that the mass scale is unchanged. A fragment ion, m_2^+, is tuned in under normal operating conditions ($V = V_0$), after which V is varied and precursors of m_2^+ are identified. If the dissociation leading to the production of m_2^+ is accompanied by the release of T eV of kinetic energy, m_2^+ ions are transmitted over a range of accelerating voltages, the limits of which can be related to the energy released[6]. For processes of interest in the work to be described below, the voltage ratios at which m_2^+ ions are transmitted by the electric sector are as follows:

Process	V/V_0
(1) $m_1^+ \rightarrow m_2^+$ (no k.e. release)	m_1/m_2
(2) $m_1^{++} \rightarrow m_2^+$ (k.e. release)	$\dfrac{m_1}{2m_2} \cdot \dfrac{1}{[1 \pm (\mu T/2Ve)^{\frac{1}{2}}]^2}$
(3) $m_1^{+++} \rightarrow m_2^{++}$ (k.e. release)	$\dfrac{2m_1}{3m_2} \cdot \dfrac{1}{[1 \pm (\mu T/3Ve)^{\frac{1}{2}}]^2}$
(4) $m_1^{+++} \rightarrow m_2^+$ (k.e. release)	$\dfrac{m_1}{3m_2} \cdot \dfrac{1}{[1 \pm (\mu T/3Ve)^{\frac{1}{2}}]^2}$

where $\mu = (m_1 - m_2)/m_2$.

Because of approximations made in deriving the expressions for V/V_0 in cases (2)–(4), the value of T, calculated in this way, is less than the true value. This may be found by evaluating T for different values of V_0 and extrapolating to the value which would be obtained at zero accelerating voltage, but this can only be done in the few cases where the intensity of the signal is sufficiently strong. In addition, if a multiply-charged ion dissociates into a heavy fragment ion and a light fragment ion, e.g. $330^{++} \rightarrow 315^+ + 15^+$, the latter will carry almost all the kinetic energy liberated in the dissociation; the peak due to the heavy ion will be broadened only very little under these circumstances so that an accurate evaluation of T is difficult, whereas the signals due to the light ion will be of low intensity and, therefore, will be difficult to locate with sufficient precision.

When no kinetic energy is released as in case (1), the plot of intensity against accelerating voltage gives a gaussian-shaped peak for a given decomposition so that the resolution of the instrument, when operated in this way, is only 50–60, based on a 10% valley definition. This has two undesirable consequences. Since the value of V/V_0 required to produce the maximum signal intensity for a given metastable transition can rarely be measured to better than $\pm$ 0·2%, difficulty may be encountered in identifying precursor ions, unequivocally, at high V/V_0 ratios, if the daughter ion is above approximately *m/e* 250. For example, for *m/e* 300 daughter ions, if $V/V_0 = 1{\cdot}700 \pm 0{\cdot}003$, the precursor ion could be *m/e* 509, 510 or 511, and the average of several readings is usually necessary before the transition can be assigned with certainty.

More serious difficulties arise if peaks due to more than one metastable transition overlap. If the peaks are of comparable intensities, a single broad peak will result, having a maximum which does not correspond to the V/V_0 ratio of either transition. Such doublets can usually be recognized from the width of the peak, and provisional assignments of metastable transitions can often be confirmed by observing peaks due to metastable transitions which occur between the electric and magnetic sectors under normal operating conditions.

If, on the other hand, one peak is much more intense than the other, the V/V_0 ratio for maximum ion current usually corresponds quite closely to that expected for the transition giving the stronger peak, and the second transition is difficult to detect with certainty since it merely leads to a slight asymmetry in the peak shape. If the neutral fragment lost contains several carbon atoms, a slightly asymmetric peak shape is, in any case, to be expected, owing to contributions from isotopic metastable peaks arising from ^{13}C isotopes. For most transitions encountered, approximate calculations show that such contributions have little effect on the V/V_0 ratio for maximum intensity but produce a slight 'tail' on the high-voltage side of the peak. Although these limitations may lead to a few low-intensity metastable transitions being overlooked, they have caused few difficulties in the present work.

3.6.3. RESULTS AND DISCUSSION

The partial mass spectrum of 9,10-diphenylanthracene, obtained using 70 V electrons, is given in *Tables 3.6.1* and *3.6.2*, and it is seen that the peak due to the molecular ion and its isotope peak are the two most intense peaks of the spectrum. The only other peaks of any prominence arising from singly-charged ions are those at *m/e* 253 and 252. The groups of peaks in the region *m/e* 125–165 are listed in *Table 3.6.2* and are thought to be due almost entirely to doubly-charged ions, since the peaks are completely absent from a 20 eV mass spectrum. The intensities in *Tables 3.6.1* and *3.6.2* are normalized to $M^+ = 100$ and $M^{++} = 100$ respectively ($M^{++} = 3{\cdot}1\%\ M^+$) and it is seen that, whereas M^+ is by far the most

Table 3.6.1. Some Singly-charged Ions in the Mass Spectrum of 9,10-diphenylanthracene

m/e	*Intensity*	*m/e*	*Intensity*
331	27·6	276	1·7
330	100·0	275	0·5
329	9·6	274	1·7
328	7·2		
327	7·5	263	0·5
326	9·0	262	0·2
325	1·3	261	0·7
324	2·9		
		254	2·2
315	1·0	253	12·6
314	1·6	252	26·4
313	3·8	251	2·8
		250	6·5
303	1·1	249	1·0
302	2·4	248	1·3
301	1·4		
300	2·9	239	0·5
299	0·6	238	0·2
298	1·4	237	0·7
		236	0·2
289	1·9	235	0·4
288	0·4		
287	1·2	226	1·6
286	0·2	225	1·0
285	0·5	224	1·4

abundant singly-charged ion, many doubly-charged fragment ions are more abundant than the doubly-charged molecular ion, the most abundant being $C_{25}H_{13}^{++}$ at *m/e* 156·5, i.e. $(M-17)^{++}$.

Table 3.6.2. Doubly-charged Ions in the Mass Spectrum of 9,10-diphenylanthracene

m/e	*Intensity*	*m/e*	*Intensity*
331/2	27	316/2	12
330/2	100	315/2	50
329/2	23	314/2	146
328/2	69	313/2	211
327/2	77	312/2	31
326/2	177	311/2	58
325/2	39		
324/2	104	304/2	13
323/2	12	303/2	31
322/2	35	302/2	108

Table 3.6.2. (continued)

m/e	*Intensity*	*m/e*	*Intensity*
301/2	63	273/2	15
300/2	204	272/2	35
299/2	38		
298/2	113	264/2	10
		263/2	33
290/2	27	262/2	15
289/2	54	261/2	30
288/2	31	260/2	10
287/2	85		
286/2	19	253/2	19
285/2	31	252/2	72
284/2	77	251/2	15
278/2	12	250/2	64
277/2	23	249/2	13
276/2	85	248/2	64
275/2	23	247/2	50
274/2	96	246/2	12

Below *m/e* 110, a large number of peaks due to triply-charged ions were observed, extending down to *m/e* 82·3, i.e. 250^{+++}. The most intense of these was at *m/e* 108·7, i.e. 325^{+++}, the intensity of which was approximately three times that of *m/e* 110 (M^{+++}) and 10% of that of M^{++}. The occurrence of these peaks together with those due to singly- and doubly-charged ions, leads to a complicated spectrum in the region of *m/e* 100, with peaks at *m/e* 99·0, 99·3, 99·5, 99·7, 100·0, 100·3, 100·5, 100·7 and 101. Although no appearance potential studies were made of triply-charged ions, peaks due to them were largely absent in a 50 eV mass spectrum.

Table 3.6.3. Metastable Transitions of Singly-charged Ions in the Mass Spectrum of 9,10-diphenylanthracene

Parent	*Daughter*	*Neutral*	*Parent*	*Daughter*	*Neutral*
$C_{26}H_{18}^+$	$C_{25}H_{15}^+$	CH_3	$C_{26}H_{17}^+$	$C_{25}H_{14}^+$	CH_3
	$C_{24}H_{15}^+$	C_2H_3		$C_{24}H_{14}^+$	C_2H_3
	$C_{23}H_{11}^+$	C_3H_7		$C_{23}H_{13}^+$	C_3H_4
	$C_{22}H_{12}^+$	C_4H_6		$C_{23}H_{10}^+$	C_3H_7
	$C_{22}H_{10}^+$	C_4H_8		$C_{22}H_{11}^+$	C_4H_6
	$C_{21}H_{11}^+$	C_5H_7		$C_{20}H_{12}^+$	C_6H_5
	$C_{21}H_9^+$	C_5H_9	$C_{26}H_{16}^+$	$C_{25}H_{13}^+$	CH_3
	$C_{20}H_{13}^+$	C_6H_5		$C_{24}H_{13}^+$	C_2H_3
	$C_{20}H_{12}^+$	C_6H_6		$C_{23}H_{12}^+$	C_3H_4
	$C_{19}H_{11}^+$	C_7H_7	$C_{26}H_{15}^+$	$C_{24}H_{12}^+$	C_2H_3

Table 3.6.3. (continued)

Parent	*Daughter*	*Neutral*	*Parent*	*Daughter*	*Neutral*
$C_{26}H_{15}^+$	$C_{22}H_9^+$	C_3H_6	$C_{24}H_{12}^+$	$C_{22}H_{10}^+$	C_2H_2
	$C_{20}H_{11}^+$	C_6H_4	$C_{23}H_{13}^+$	$C_{21}H_{11}^+$	C_2H_2
$C_{26}H_{14}^+$	$C_{24}H_{12}^+$	C_2H_2	$C_{23}H_{12}^+$	$C_{21}H_{10}^+$	C_2H_2
$C_{26}H_{12}^+$	$C_{20}H_{10}^+$	C_6H_2	$C_{23}H_{11}^+$	$C_{21}H_9^+$	C_2H_2
$C_{26}H_{11}^+$	$C_{24}H_{10}^+$	C_2H	$C_{22}H_{11}^+$	$C_{20}H_0^+$	C_2H_2
	$C_{20}H_9^+$	C_6H_2	$C_{22}H_{10}^+$	$C_{20}H_8^+$	C_2H_2
$C_{26}H_{10}^+$	$C_{24}H_9^+$	C_2H	$C_{21}H_{12}^+$	$C_{19}H_{10}^+$	C_2H_2
	$C_{20}H_8^+$	C_6H_2	$C_{21}H_{11}^+$	$C_{19}H_9^+$	C_2H_2
$C_{25}H_{16}^+$	$C_{24}H_{13}^+$	CH_3	$C_{21}H_9^+$	$C_{19}H_7^+$	C_2H_2
$C_{25}H_{15}^+$	$C_{24}H_{12}^+$	CH_3	$C_{20}H_{14}^+$	$C_{19}H_{11}^+$	CH_3
	$C_{23}H_{13}^+$	C_2H_2	$C_{20}H_{13}^+$	$C_{19}H_{10}^+$	CH_3
$C_{25}H_{14}^+$	$C_{23}H_{12}^+$	C_2H_2	$C_{20}H_{12}^+$	$C_{19}H_9^+$	CH_3
$C_{25}H_{13}^+$	$C_{23}H_{11}^+$	C_2H_2		$C_{18}H_{10}^+$	C_2H_2
$C_{24}H_{14}^+$	$C_{22}H_{12}^+$	C_2H_2	$C_{20}H_{11}^+$	$C_{18}H_9^+$	C_2H_2
$C_{24}H_{13}^+$	$C_{22}H_{11}^+$	C_2H_2	$C_{20}H_{10}^+$	$C_{18}H_8^+$	C_2H_2

The 70 eV low resolution mass spectrum contains only one prominent metastable peak; this peak is rather broad and unsymmetrical and closer examination shows that it is a composite peak with maxima at *m/e* 192·5 and *m/e* 194·0, the latter being slightly more intense. These can be ascribed to the two reactions:

$$\underset{(330^+)}{C_{26}H_{18}^+} \rightarrow \underset{(252^+)}{C_{20}H_{12}^+} \quad + \quad C_6H_6 \qquad \ldots (3.6.1)$$

$$\rightarrow \underset{(253^+)}{C_{20}H_{13}^+} \quad + \quad C_6H_5 \qquad \ldots (3.6.2)$$

which provide routes for the two most important singly-charged fragment ions. However, the metastable peak arising from the process:

$$\underset{(329^+)}{C_{26}H_{17}^+} \rightarrow \underset{(252^+)}{C_{20}H_{12}^+} \quad + \quad C_6H_5 \qquad \ldots (3.6.3)$$

would lie at *m/e* 193·1 and it is impossible to say, from these observations alone, whether or not this process occurs.

A very large number of metastable transitions were observed due to decompositions which occurred between the source and the electric sector. The most prominent of those detected for singly-charged ions are listed in *Table 3.6.3.* When tuned to the *m/e* 252 peak, a maximum in the ion current was observed with an accelerating voltage ratio of $V/V_0 = 1{\cdot}3074 \pm 0{\cdot}0005$, this being the average of eight determinations. The occurrence of the reactions of equations 3.6.1 and 3.6.3 would lead to expected voltage ratios of 1·3096 and 1·3056 respectively, so that the observed maximum suggests that the $C_{20}H_{12}^+$ ion is formed by both routes. It is unlikely that the discrepancy can be ascribed to experimental error, since similar observations on the *m/e* 253 peak led to $V/V_0 = 1{\cdot}3048 \pm 0{\cdot}0005$, in good agreement with the expected value of 1·3044 calculated for the reaction of equation 3.6.2.

In general, the results in *Table 3.6.3* show that CH_3 and C_2H_2 are the neutral fragments most frequently lost by singly-charged ions, with the exception of the fragments lost by C_{26} ions. The molecular ion appears to lose a C_2H_3 fragment:

$$C_{26}H_{18}^+ \rightarrow C_{24}H_{15}^+ \quad + \quad C_2H_3 \qquad \ldots (3.6.4)$$

and this also seems to be the main C_2 fragment lost by $(M-1)^+$–$(M-3)^+$. Although it is possible that the reaction may occur in two steps:

$$C_{26}H_{18}^+ \xrightarrow{-H} C_{24}H_{17}^+ \xrightarrow{-C_2H_2} C_{24}H_{15}^+ \qquad \ldots (3.6.5)$$

the peaks due to the metastable transitions are fairly sharp and appear to indicate that the C_2H_3 entity is lost in one step.

Some 40 metastable transitions of doubly-charged species were observed and in almost all cases, CH_3^+ or $C_2H_3^+$ was the lighter fragment produced. Since these ions carry almost all the energy liberated in the decompositions, the peaks due to the heavier fragment ions were broadened very little and the energies released in most processes could not be evaluated. Low intensity signals, indicating the loss of C_3 fragment ions, were also observed in a number of cases. The doubly-charged molecular ion M^{++} $(M-1)^{++}$ and $(M-2)^{++}$ were all observed to lose $C_6H_6^+$, and, in the case of the molecular ion, the kinetic energy released in the reaction was estimated to be $1{\cdot}6 \pm 0{\cdot}2$ eV.

A similar search was made for metastable transitions of triply-charged ions, and rather more than 30 were observed. The pattern was very similar to that found for doubly-charged ions, with CH_3^+ and $C_2H_3^+$ again being the lighter ions produced in the vast majority of decompositions, together with a few cases in which $C_3H_3^+$ is lost. The triply-charged molecular ion again loses $C_6H_6^+$ to form $C_{20}H_{12}^{++}$ and, in this case, the kinetic energy released in the decomposition is $2{\cdot}4 \pm 0{\cdot}3$ eV.

Intensity variations, both of peaks observed in the normal mass spectrum and of the signals obtained for the metastable transitions, call for some comment. Examination of *Table 3.6.1* shows that when the fragment ions contain an even number of carbon atoms (C_{26}, C_{24}, C_{22}, C_{20} and C_{18}), it is the even mass ions which tend to be the most abundant, whereas odd mass ions tend to be the most abundant when fragment ions contain an odd number of carbon atoms (C_{25}, C_{23}, C_{21} and C_{19}). Once the pattern has been established by the higher mass ions, the successive loss of CH_3 or C_2H_2 fragments would reproduce the pattern at lower mass numbers, and it is these fragments which are most commonly lost as the data in *Table 3.6.3* indicate.

A very similar alternation in intensities is found for doubly-charged ions as the data in *Table 3.6.2.* shows This suggests that a similar pattern is established for doubly-charged fragment ions and is reproduced by the successive loss of neutral CH_3 or C_2H_2 fragments. About 25 metastable transitions of the type $m_1^{++} \rightarrow m_2^{++} + (m_1 - m_2)$ were observed, and CH_3 and C_2H_2 were by far the most commonly lost fragments, although C_2H_4, C_3H_4 and C_6H_6 were lost by the doubly-charged molecular ion.

It has been pointed out that in a series of metastable transitions given by doubly-charged C_6 species, it is the even-mass ions such as $C_6H_6^{++}$, $C_6H_8^{++}$ and $C_6H_{10}^{++}$ which give rise to the more intense metastable transitions[7]. This can be rationalized by observing that these are all even-electron species which may be expected to be more stable than odd-electron species so that a greater fraction of such ions will decompose with rate constants of 10^5–10^6 s^{-1} than is the case for odd-electron species. A search for a similar trend was made in the present work, but none was found.

The many different types of metastable transition, observed in the mass spectrum of this compound, suggest that although many fragment ions are formed by the decomposition of the singly-charged molecular ion, a significant fraction are formed by the fragmentation of multiply-charged species. For example, a sequence such as:

$$C_{26}H_{17}^{+++} \xrightarrow{-CH_3^+} C_{25}H_{14}^{++} \xrightarrow{-C_2H_3^+} C_{23}H_{11}^{+} \ldots \quad (3.6.6)$$

could well occur, also others in which neutral fragments are lost by multiply-charged ions before a singly-charged fragment ion is lost. Furthermore, since light fragment ions such as CH_3^+ and $C_2H_3^+$ are formed in such a large number of processes, the kinetic energies carried by these ions are likely to vary very considerably, depending on the particular reaction sequence leading to their production.

We thank AEI Ltd. for a maintenance grant to A. F. Whiting.

REFERENCES

[1] WACKS, M. E. and DIBELER, V. H. *J. chem. Phys.* 31 (1959) 1557

[2] MEYERSON, S. and VAN DER HAAR, R. *J. chem. Phys.* 37 (1962) 2458

[3] HIGGINS, W. and JENNINGS, K. R. *Trans. Faraday Soc.* 62 (1966) 96

[4] JENNINGS, K. R. *Z. Naturf.* 22A (1967) 454

[5] JENNINGS, K. R., *Newer Physical Methods in Organic Structure Determination*, United Trade Press, Oxford, 1967

[6] BARBER, M., JENNINGS, K. R. and RHODES, R. *Z. Naturf.* 22A (1967) 15

[7] BARBER, M. and JENNINGS, K. R. *Am. Soc. Test. Mater. 15th A. Conf. Mass Spectrom. Allied Topics*, Denver (1967) Z. *Naturf.* (in press)

SESSION 4

CHEMICAL APPLICATIONS II

4.1

USE OF MULTIPLET PEAKS IN THE EXAMINATION OF HIGH MOLECULAR WEIGHT PETROLEUM FRACTIONS

W. K. Reid and W. L. Mead

4.1.1. INTRODUCTION

THE petroleum industry has always been at the forefront in applying instrumental analytical techniques. As soon as suitable instrumentation has become available, the petroleum analyst has quickly turned it to his advantage in the study of complex mixtures. Since the 1940s, from work of Washburn *et al.*[1–4] on gases and light distillates, petroleum chemists have been utilizing the fragmentation pattern of hydrocarbon molecules in the mass spectrometer to provide analyses of a wide range of petroleum fractions. It is surprising that the organic chemist did not fully realize the potential of mass spectrometry until the 1950s, even when the early equipment was often suitable for his problems. Once this realization had dawned on the organic chemist, he became impatient for more detailed information from the mass spectrum and the past 5 years has seen a staggering development in the field of organic structure determination, pioneered principally by Beynon, Biemann, Djerassi, McLafferty and Reed *et al.*

Perhaps the most important instrumental advance in the past few years has been the development of high resolution mass spectrometry, principally by Beynon *et al.*, and again the petroleum chemist has not been slow in appreciating the value of this development. This paper gives a brief review of high resolution mass spectrometry in the petroleum field and discusses some applications of current interest.

4.1.2. HYDROCARBON TYPE ANALYSIS

4.1.2.1. The 'Fragment' Peak Method

Conventional, low resolution mass spectrometry has already found wide application in the petroleum industry. For many years, this technique has provided hydrocarbon type analyses of petroleum fractions for refinery and process control, in petrochemical feed-stocks and general research into the composition of petroleum. The method employed was first introduced by Brown *et al.*[5] for gasolines. Briefly, the method involves measurement and summation of the peak heights of certain series of peaks in the spectrum which are characteristic of the various hydrocarbon types, application of these summations to a set of simultaneous equations, which correct for contributions from one series to the others, and solution of the equations to give the amount of each hydrocarbon type present. This is, of course, a simplified statement of the method and other factors, for example isotopic contributions, calibration data, must also be considered.

However, useful though these analyses are, the possible presence of non-hydrocarbon peaks due to sulphur, nitrogen and oxygen compounds has had to be ignored, since, under low resolution conditions, these peaks are overlapped by hydrocarbon peaks. The effect of these compounds is usually small when dealing with low molecular weight petroleum fractions, but if any separation procedure has been used that concentrates the hetero-atom compounds, the analysis on a hydrocarbon basis may be completely misleading. This has been discussed in some detail elsewhere[6].

High resolution mass spectrometers have the ability to resolve the majority of these overlapping peaks into the individual peaks due to the hydrocarbons and the various hetero-atom compounds that are present, and it is obvious that a major advance would be achieved if these multiplet peaks could be used to provide a comprehensive analysis, not only of the hydrocarbons, but of all compound types present. There are three main difficulties. There is a lack of calibration data from pure compounds, particularly non-hydrocarbons at high molecular weights; the choice of peak series which best represents a particular compound type is difficult; and there is the problem of the large amount of data that would have to be collected and processed. This may be solved when data handling systems based on, for example, the 'element mapping' method of Biemann[7] become more generally available for complex systems.

Recently, however, a paper has been presented[7a] in which the 'fragment peak' method has been combined with high resolution to give group type analyses of high boiling petroleum stocks. This method does not require prior fractionation of the sample. The authors found that data from the AEI MS9 are sufficiently reproducible for this purpose and determined 19 component types, including seven saturate hydrocarbons with zero to six rings, nine aromatics with one to four rings and three aromatic sulphur compound types.

4.1.2.2. Low Voltage Method

Although high resolution conditions have not yet been applied in the 'fragment peak' method, to much extent, the method of low voltage analysis pioneered by Field and Hastings[8] and developed by Lumpkin *et al.*[9–11], for hydrocarbon type analysis of aromatic hydrocarbons, has been applied by Lumpkin[12], Reid, Mead and Bowen[6] and Johnson and Aczel[13, 13a] at high resolution using AEI MS9 mass spectrometers.

Table 4.1.1. Summary Analysis of a Petroleum Fraction (Lumpkin)

Mass series		*Weight* (%)
C_nH_{2n-8}	Indanothiophenes	4·2
C_nH_{2n-10}	Indenothiophenes	1·8
	Indenes	0·4
C_nH_{2n-12}	Dibenzothiophenes	39·7
	Naphthalenes	2·3
C_nH_{2n-14}	Acenaphthenothiophenes	1·5
	Dibenzofurans	2·4
	Acenaphthenes	7·9
C_nH_{2n-15}	Carbazoles	2·8
C_nH_{2n-16}	Fluorenones	1·1
	Fluorenes	13·3
C_nH_{2n-18}	Phenanthrenes	16·8
C_nH_{2n-20}	Benzodithiophenes	2·8
	Cyclopentaphenanthrenes	3·0
		100·0

The technique of low voltage analysis is useful in that substantially only the parent ions of molecules are observed in the spectrum, though the method is limited to unsaturated species. This method has been well developed at low resolution by Lumpkin *et al.*[10, 11]. A very useful feature of their technique is a method of obtaining calibration data at high molecular weights, where pure materials are usually unobtainable, by extrapolating from low molecular weights. To do this, relative sensitivities of pure components are plotted against reciprocal molecular weight and straight-line extrapolations give calibration data at higher molecular weights.

Lumpkin[12] and workers at the BP Research Centre[6] explored the possibilities of combining high resolution with low electron voltage to enable sulphur compounds and hydrocarbons to be estimated. Lumpkin examined the trinuclear aromatic portion of a coker gas oil and, using accurate mass measurement by the MS9

Table 4.1.2. Analysis of Aromatic Plus Sulphur Fraction from Gas Oil at Low Voltage and High Resolution. (*From* W. K. Reid, W. L. Mead and K. M. Bowen[6], *by courtesy of the* Institute of Petroleum)

Compound type	*Weight* (%)	*Compound type*	*Weight* (%)
C_nH_{2n-6}	14·1	C_nH_{2n-14}	6·5
$C_nH_{2n-10}S$	5·3	$C_nH_{2n-18}S$	2·7
C_nH_{2n-8}	7·3	C_nH_{2n-16}	10·8
$C_nH_{2n-12}S$	2·8	$C_nH_{2n-20}S$	0·6
C_nH_{2n-10}	4·3	C_nH_{2n-18}	9·8
C_nH_{2n-12}	5·5	$C_nH_{2n-22}S$	0·2
$C_nH_{2n-16}S$	30·1		

Average molecular weight of sulphur compounds = 250
Sulphur content (calculated) = 5·2% weight
Sulphur content (experimental) = 4·6% weight

peak matching technique[14], obtained an analysis in terms of the compound types shown in *Table 4.1.1.*, in some cases using mass measurements on fragment peaks at high electron voltage to identify the components. Reid, Mead and Bowen[6] compared the fragment peak and low electron voltage methods in some detail, on the AEI MS2 and MS9 mass spectrometers and, using Lumpkin's extrapolation method developed calibration data for some hydrocarbon and sulphur compounds (*Figures 4.1.1* and *4.1.2*). From these, analyses of the aromatic plus sulphur fraction of gas oils are obtainable (*Table 4.1.2*), in terms of the amounts of sulphur and hydrocarbon compounds, and the distribution of both these classes of compounds can be plotted (*Figure 4.1.3*).

However, although such results represent a considerable step forward, the peak matching technique is slow, particularly when a large number of peaks have to be accurately mass measured. Recently, Johnson and Aczel[13] have shown that it is possible to use chart measurements of high resolution spectra of petroleum fractions and that the accuracy of mass measurement approaches that obtained by the peak matching technique, even though only a ruler and a simple magnifying glass are used. However, it should be stated that any major advance in the routine analysis of petroleum fractions employing high resolution mass spectrometry, must surely depend on two factors. Data handling equipment, possibly with an

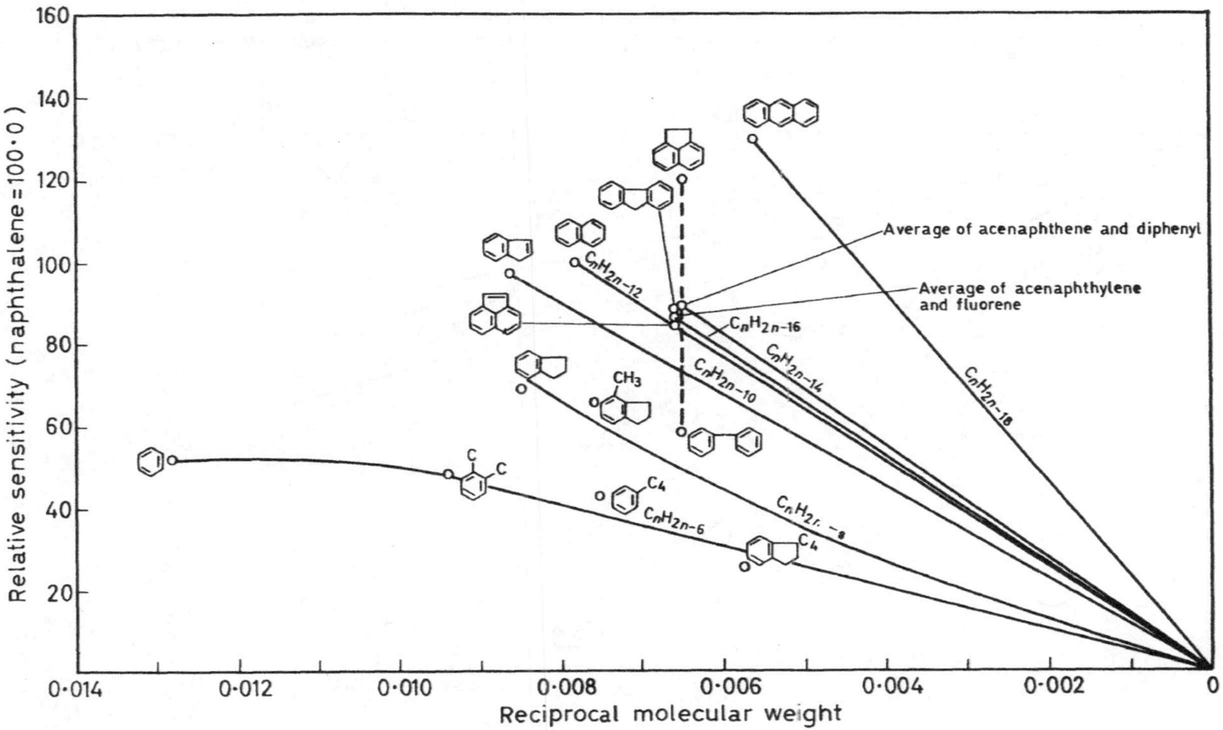

Figure 4.1.1. Low voltage calibration graph for some hydrocarbon types; relative sensitivity plotted against reciprocal molecular weight with naphthalene having standard sensitivity of 100. (*From* W. K. Reid, W. L. Mead and K. M. Bowen[6], *by courtesy of the* Institute of Petroleum)

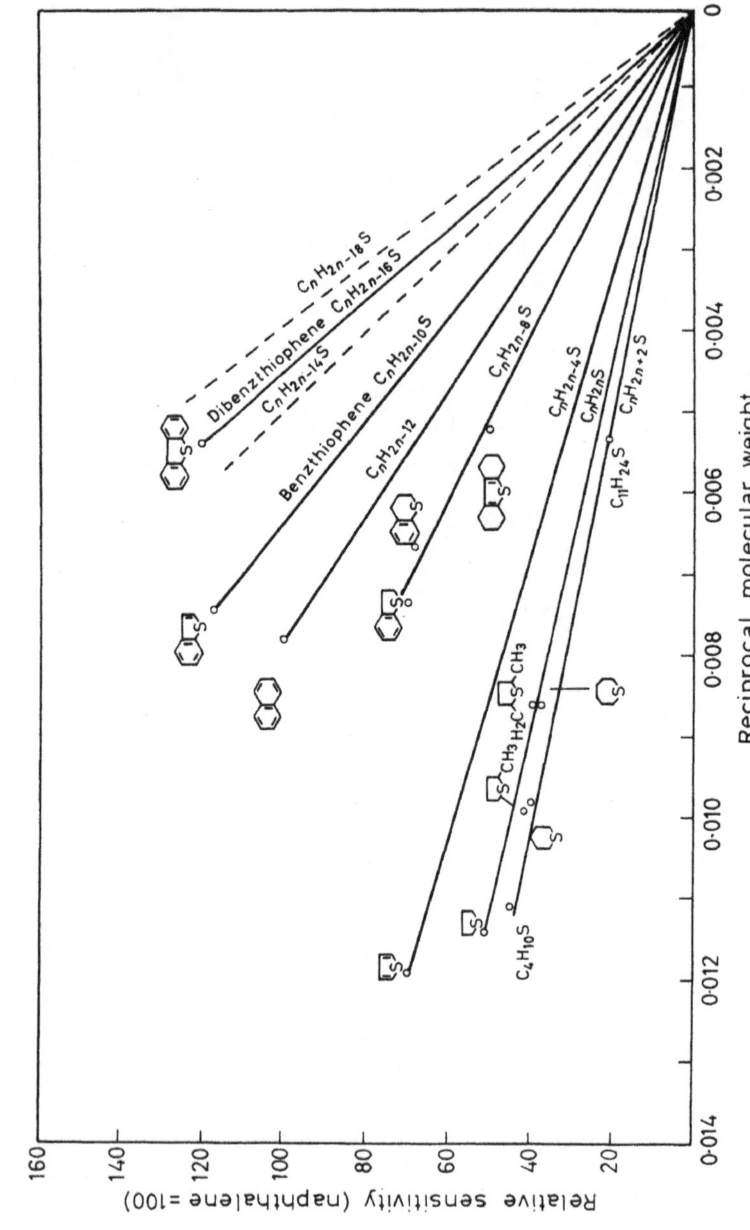

Figure 4.1.2. Low voltage calibration graph for some sulphur compound types. (From W. K. Reid, W. L. Mead and K. M. Bowen[6], *by courtesy of the* Institute of Petroleum)

on-line computer, must be available to record peak intensities, make mass measurements and interpret the data automatically and better calibration data should be available.

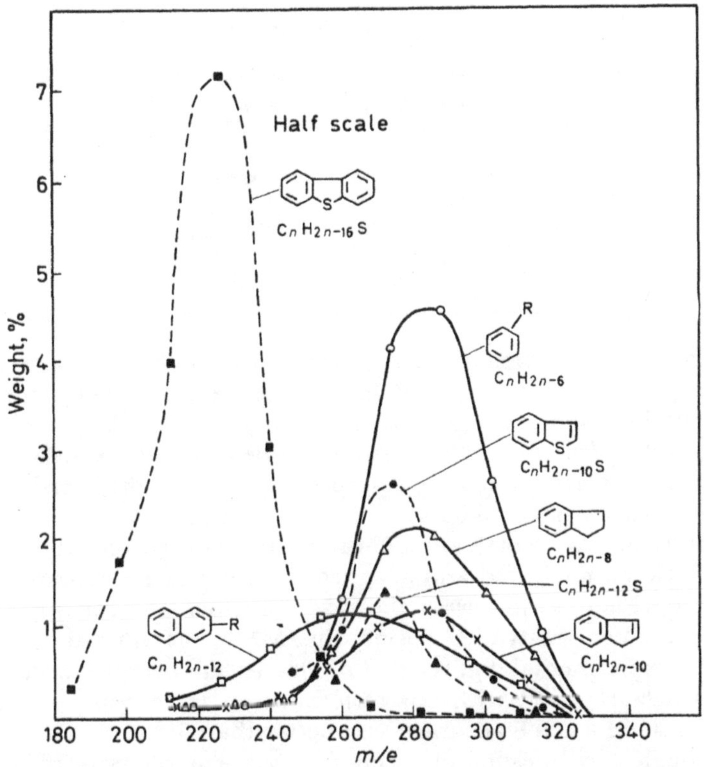

Figure 4.1.3. Quantitative distributions of some hydrocarbons and sulphur compound types obtained from the MS9 *spectrum of an* A + S *portion of a Kharg gas oil* (343–371°C) *at low voltage.* (*From* W. K. Reid, W. L. Mead and K. M. Bowen[6], *by courtesy of the* Institute of Petroleum)

4.1.3. ACCURACY OF MASS MEASUREMENT

The accuracy of mass measurement is a critical factor in the assignment of ionic formulae. In this respect, the organic chemist, when dealing with a pure compound, is usually satisfied by an accuracy of approximately 3–5 p.p.m. or often up to 10 p.p.m., depending on the nature and his preknowledge of the sample.

In complex mixtures such as petroleum fractions, it may be said that the petroleum analyst is never really satisfied with the limits of accuracy obtainable. *Table 4.1.3* lists some of the common doublets encountered in the examination of petroleum fractions and it is seen that some doublets, particularly the $S–C_3H_{-4}$ doublet, involve only very small mass differences.

Table 4.1.3. Common Doublets in High Resolution Mass Spectra

Doublets	*Mass difference,* Δm (u)	*Resolution required at mass* 200, $200/\Delta m$
$H_{12}–C$	0·0939	2130
$CH–C^{13}$	0·0045	44,000
$CH_2–N$	0·0126	15,900
$CH_4–O$	0·0364	5500
$C_2H_8–S$	0·0905	2210
$S–C_3H_{-4}$	0·0034	59,000

As the molecular weight range of the petroleum fraction increases, so does the accuracy required in the mass measurement but the certainty in assignment decreases. Thus, it is unlikely that the higher boiling fractions of petroleum will be dealt with satisfactorily until a new generation of mass spectrometers is produced, with resolution upwards of 50,000.

Tunnicliff, Wadsworth and Schissler[21] have stated that the agreement between measured and assigned mass must be approximately 0·0002 u without considering the mass at which the measurement is made. Van Katwijk[22], using the MS9, has quoted a more realistic experimental figure of 3 p.p.m. as an average accuracy. This is in agreement with our own work on trace components in waxes[19], in which the accuracy was better than 5 p.p.m.

Although the question of accuracy is a troublesome one, the combination of other spectroscopic and separation techniques can often be used successfully in narrowing the number of possibilities that fall within the accuracy limits. Interesting reviews have recently been given on this subject[23, 24] at the World Petroleum Congress.

4.1.4. ION IDENTIFICATION IN HIGH RESOLUTION MASS SPECTRA

Once an accurate mass measurement has been obtained on a particular ion in a mass spectrum, e.g. from comparison with an internal standard, the problem of assigning a molecular formula to the ion then arises. If the ion is from a pure compound, the organic chemist can usually obtain a molecular formula by looking up the accurate mass in a set of tables, such as those devised by Beynon[15]. In this task, he is assisted usually by some preknowledge of the sample source and often the problem reduces to one of a simple nature, e.g. does the compound contain two or three atoms of sulphur?

However, as was realized by Carlson *et al.*[16], ion identification in high resolution mass spectra of petroleum fractions, can be difficult. The number and type of hetero-atoms is uncertain, unresolved doublets may occur and the spectrum, in which fragment and parent

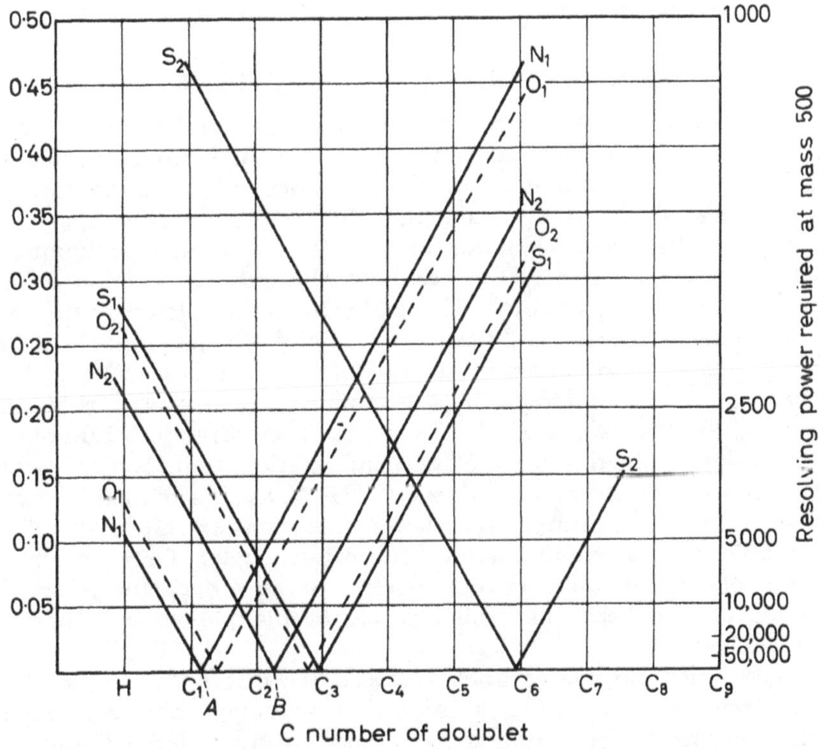

Figure 4.1.4. Δm *for the doublet series* N, N_2, S, S_2, O_1 *and* O_2*–hydrocarbon,.* (*After* W. K. Reid, W. L. Mead and A. R. West[18], *by courtesy of* Analytical Chemistry)

ions may overlap, is exceedingly complex. These workers measured the difference between an unknown ion and a reference hydrocarbon ion to obtain the assignment. Kendrick[17] also developed a method for ion identification based on a mass standard of $CH_2 = 14{\cdot}000000$

which is particularly useful for petroleum work since homologous series give the same mass defect.

In our laboratories, a method has been developed[18] which is particularly useful when dealing with complex mixtures and which has been used extensively in our work in this field[6, 19, 20]. In this method the values of the doublet differences between hetero-atom combinations and the corresponding hydrocarbons, e.g. S_2–C_3H_{28}, S_2–C_4H_{16} are plotted against carbon number (*Figure 4.1.4.*) The list of doublets can be quickly extended since all the lines are parallel and thus this technique is extremely useful for assessment of doublet values of nitrogen, sulphur and oxygen occurring together, even in unusual combinations.

4.1.5. SEPARATION TECHNIQUES

High resolution mass spectrometry is a new technique in the petroleum industry but has already proved valuable in its application. However, future developments may depend on preseparation techniques which would make the high resolution mass spectrometric results easier to obtain and interpret. Simple techniques such as concentration of the aromatic plus sulphur fraction on a silica gel column, prior to MS9 analysis of a petroleum distillate, have already been widely used[6] and Lumpkin[12] has used distillation, silica gel chromatography, and thermal diffusion followed by alumina gel chromatography, to obtain a trinuclear aromatic fraction from a coker gas oil. From this type of fraction, Lumpkin obviously obtained more detailed information than would have been possible on the original distillate. However, it is our experience that a simply-separated petroleum fraction can often contain valuable internal standards for mass measurement, for example a series of intense alkylbenzene parent peaks are useful internal standards 14 u apart, and this idea has been exploited by Johnson and Aczel[13].

Gas–liquid chromatography has also recently been coupled to high resolution instruments using fast scanning, magnetic tape recording and data handling techniques by computer[25, 26] but as yet no major attack has been made on the problem of petroleum mixtures using this powerful combination. It appears that this will be an important technique when the required instrumentation becomes more generally available. However, many workers in the petroleum industry are now attaching g.l.c. columns to their low resolution mass spectrometer, e.g. Dorsey[27] has already identified 18 C_9 isomers in the saturate fraction of a reformate.

The technique of gel permeation chromatography, originally a biochemical method for separation of high molecular weight natural products, suggests interesting possibilities of fractionation in the higher boiling and residual fractions from crude oil, because of its ability to separate fractions on molecular size. For example, Altgelt[28] has already fractionated asphaltenes and obtained a porphyrin concentrate.

4.1.6. GENERAL APPLICATIONS

4.1.6.1. Structure and Origin of Petroleum

High resolution mass spectrometry has applications in research studies on the structure and origin of petroleum. Hills and Whitehead[29, 29a] have established the presence of pentacyclic triterpanes in petroleum, using a combination of high resolution mass spectrometry, X-ray crystallography and nuclear magnetic resonance. This work may well be useful in determining whether the crude oil, from which individual compounds are isolated, is of marine or terrestrial origin. Baker[30] has used an MS9 to obtain mass measurements, and hence molecular formulae, for a series of porphyrin fractions extracted from asphaltenes, and considers his results to have significance on the geochemistry of the formation of petroporphyrins. In this study, Baker employed low electron voltages to obtain distributions of the porphyrins present.

4.1.6.2. Trace Components

By high resolution mass spectrometry, it is often possible to detect trace amounts of materials in a petroleum mixture, even without preseparation, provided that the mass difference of the component peak is sufficient to resolve it from peaks of the same nominal mass. For example, small amounts of oxygen compounds have been observed in waxes and distributions of compound types of minor components obtained[19].

4.1.7. RECENT INVESTIGATIONS

Our recent work, employing high resolution, has been mainly in three specific regions—field ionization, development of analysis of lubricating oils and a study of some thermally degraded asphaltenes. These studies are not complete but they serve to indicate the analytical potential of the technique in the petroleum field.

4.1.7.1. Field Ionization

Work has been conducted in our laboratories recently in co-operation with AEI Manchester, who have developed a combined field ionization–electron bombardment source for attachment to their MS9 and MS12 mass spectrometers. This preliminary work has been reported[32] and in this paper some of the more important applications to the petroleum industry will be mentioned.

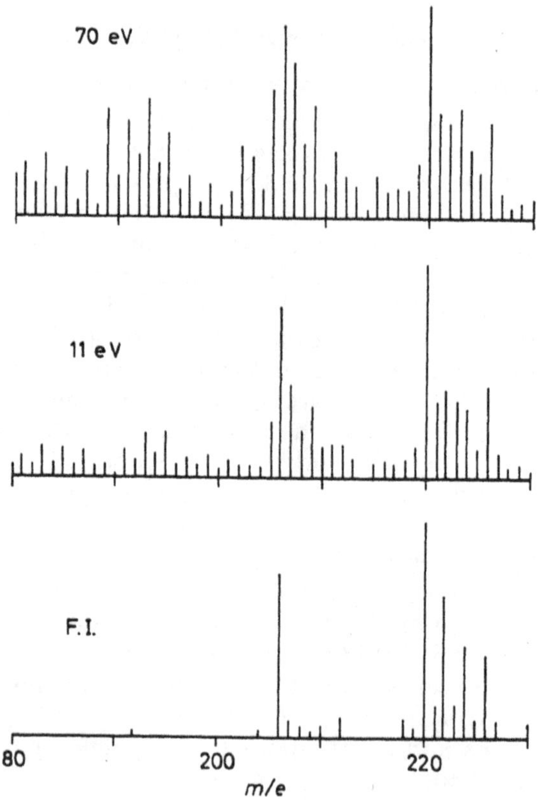

Figure 4.1.5. A + S *heavy gas oil spectra*

Briefly, field ionization spectra, an extensive study of which has been made by Beckey, are produced by applying a high voltage to a fine edge or tip. The technique has the advantage of producing spectra that consist of essentially only parent ions, both for saturate and unsaturate materials. Thus, just as low electron voltage conditions can be used for the analysis of aromatic plus sulphur fractions, field ionization will have important applications in the analysis of saturate fractions. Beckey *et al.*[33] have demonstrated the usefulness of the technique in the analysis of gasolines. Wanless[34] has reported some useful data on a large number of compounds relevant to the petroleum industry.

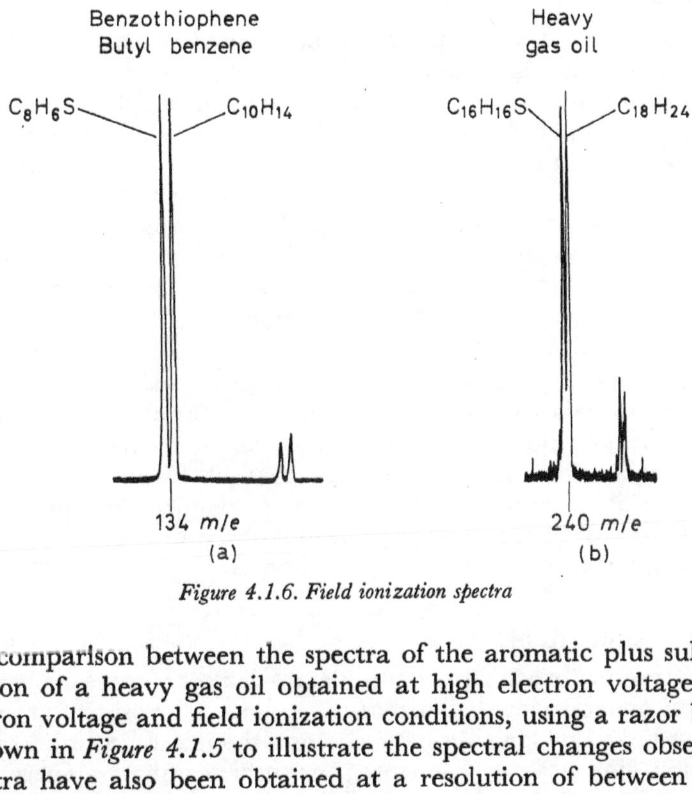

Figure 4.1.6. Field ionization spectra

A comparison between the spectra of the aromatic plus sulphur fraction of a heavy gas oil obtained at high electron voltage, low electron voltage and field ionization conditions, using a razor blade is shown in *Figure 4.1.5* to illustrate the spectral changes observed. Spectra have also been obtained at a resolution of between 2000 and 4000 under field ionization conditions, and a doublet in a gas oil spectrum is shown in *Figure 4.1.6.* Very few laboratories are exploring this technique in the petroleum field at present, but from the early results obtained, it appears that field ionization mass spectrometry will become an important addition to the petroleum analysts' techniques.

4.1.7.2. Lubricating Oil Analysis

The available 'fragment peak' and low voltage methods of hydrocarbon type analysis, which are applicable, in general, to the gas oil boiling range, are often quite useless in studying lubricating oil fractions since many more compound types are present, including both hydrocarbons and non-hydrocarbons, for which calibration data are not available. Keeping in mind the fact that any analysis should eventually be on a routine basis and that data handling systems should be available in the near future, we are attempting to devise a system of analysis for lubricating oil based on the 'fragment peak' method.

In this work perfluorokerosine is being used as an internal mass standard. Perfluorokerosine has the advantage of giving a series of known peaks, well separated from the peaks normally found in the mass spectra of petroleum fractions. The spectrum of the lubricating oil is obtained at a resolution sufficient only to split the most common doublets, $C–H_{12}$ (0·0939), $S–C_2H_4$ (0·0905) and $O–CH_4$ (0·0364). Any attempts to split closer doublets would require much higher resolution and slower scan speed and thus would not be convenient for routine application. The method envisaged uses chart distances to obtain mass measurements in the way employed by Johnson and Aczel[13].

Fractionation of the lubricating oil aromatic plus sulphur fraction is being undertaken by thermal diffusion with the object of obtaining calibration data on narrow cuts. However, some basic rethinking will be required in the choice of peaks used in the peak summations characteristic of compound types since, instead of the nine hydrocarbon types determined at low resolution, the problem will be one of dealing with 27 or more compound types.

4.1.7.3. Asphaltenes

The structure of asphaltic molecules is a matter of some conjecture, the most general view being that the molecules consist of aromatic portions linked by long aliphatic chains although some authors have suggested linkages by condensed naphthene rings. Clerc and O'Neal[31] have obtained mass spectra of an asphalt on a low resolution instrument, by placing the sample on a heater in the ion source. However, the information was obtained from the spectra merely by proposing ion formulae to fit the nominal *m/e* ratios of the peaks obtained. The spectra of asphaltenes on a high resolution instrument, using a direct-probe insertion system, have already been studied[20], mass measurements being used to identify some of the ions present although the spectra were extremely complex (*Figure 4.1.7*).

Recently, some thermally degraded samples of asphaltenes have been studied on the MS9 in our laboratories and these have indicated some of the types of structure present in the original molecule.

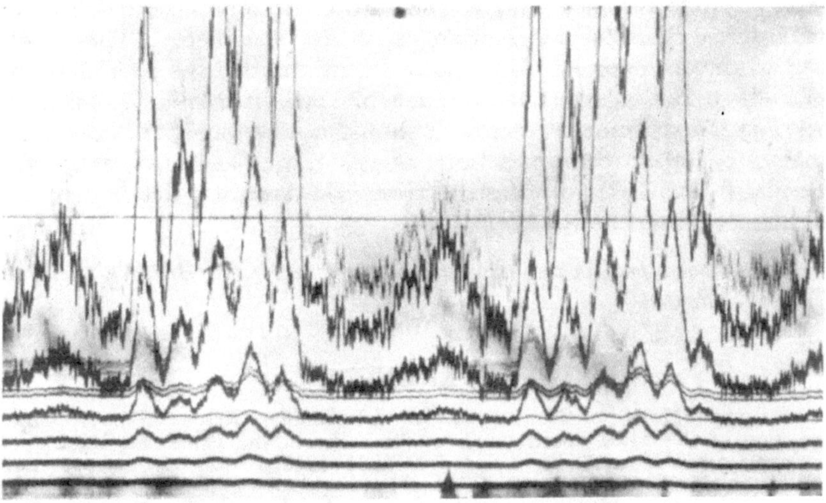

Figure 4.1.7. Portion of mass spectrum of an asphaltene

The asphaltene was thermally degraded in the absence of air, fractionated by chromatography into saturate, intermediate and aromatic fractions and the samples examined by high resolution mass spectrometry on the direct probe inlet system. The intermediate fraction was shown, by mass spectrometry, to contain mainly a C_nH_{2n-6} hydrocarbon series, ranging from C_{37} to C_{56}, which could have been tetracycloparaffins, tetraolefins, etc., or possibly alkylbenzenes. Infra-red (i.r.) and nuclear magnetic resonance (n.m.r.) were able to confirm that these were olefinic materials.

The aromatic portion was found to fractionate as the temperature of the ion source increased and three different mass spectra, corresponding to materials containing large amounts of hetero-atom aromatics, were obtained. One of the fractions was found to give multiplet peaks with five resolved peaks per mass number and, from chart measurements, distributions of certain possible compound types could be drawn. Asphaltenes will present a considerable chemical problem for many years but it may be expected that high resolution mass spectrometry will have an important part to play in the elucidation of their structure.

4.1.8. SUMMARY

High resolution mass spectrometry can make available a wealth of information on the study of petroleum fractions. However, pre-separation techniques will, in the future, become essential, if the maximum possible information is to be obtained. The most important development in the near future may be the combination of high resolution instrumentation with data handling systems and automatic mass measurement. It should not be forgotten, however, that often other techniques such as i.r., n.m.r. and u.v., etc., can be useful ancillaries to mass spectrometric data in a detailed study of petroleum fractions.

Permission to publish this paper has been given by The British Petroleum Company Limited.

REFERENCES

[1] WASHBURN, H. W. and HOOVER, H. JR., 'A Preliminary Report on the Application of the Mass Spectrometer to Problems in the Petroleum Industry'. *Am. Inst. Min. metall. Engrs Tech. Publ.* 1205 (1940)

[2] WASHBURN, H. W., TAYLOR, D. D., HOSKINS, E. E. and LANGMUIR, R. V. 'Applications of Mass Spectrometry to Analysis of Hydrocarbon Gas Mixtures' CNGA Meeting (1940)

[3] WASHBURN, H. W., WILEY, H. F. and ROCK, S. M. *Ind. Engng Chem. analyt. Edn.* 15 (1943) 541

[4] WASHBURN, H. W., WILEY, H. F., ROCK, S. M. and BERRY, C. E. *Ind. Engng Chem. analyt. Edn.* 17 (1945) 74

[5] BROWN, R. A., MELPOLDER, F. W. and YOUNG, W. S. *Petrol. Process.* 7 (1952) 204

[6] REID, W. K., MEAD, W. L. and BOWEN, K. M., *Advances in Mass Spectrometry*, Vol. 3, W. L. MEAD (Ed.), Institute of Petroleum, London, 1966

[7] BIEMANN, K. *Pure appl. Chem.* 9 (1964) 95

[7a] GALLEGOS, E. J., GREEN, J. W., LINDEMAN, L. P., LE TOURNEAU, R. L. and TEETER, R. M. 'Petroleum Group Type Analysis by High Resolution Mass Spectrometry'. *Am. chem. Soc. 153rd Meet.*, Florida (1967)

[8] FIELD, F. H. and HASTINGS, S. H. *Analyt. Chem.* 28 (1956) 1248

[9] LUMPKIN, H. E., THOMAS, B. W. and ELLIOT, A. *Analyt. Chem.* 24 (1952) 1389

[10] LUMPKIN, H. E. *Analyt. Chem.* 30 (1958) 321

[11] LUMPKIN, H. E. and ACZEL, T. *Analyt. Chem.* 36 (1964) 181

[12] LUMPKIN, H. E. *Analyt. Chem.* 36 (1964) 2399

[13] JOHNSON, B. H. and ACZEL, T. *Analyt. Chem.* 39 (1967) 682

[13a] ACZEL, T. and JOHNSON, B. H. 'High Resolution Mass Spectral Analysis of Complex Petroleum Products and their Gas Chromatographic Distillate Fractions'. *Am. chem. Soc. 153rd Meet.*, Florida (1967)

[14] QUISENBERRY, K. S., SCOLMAN, T. T. and NIER, A. O. *Phys. Rev.* 102 (1956) 1071

[15] BEYNON, J. H., *Mass Spectrometry and its Applications to Organic Chemistry*, Elsevier, Amsterdam, 1960

[16] CARLSON, E. G., PAULISSON, G. T., HUNT, R. H. and O'NEAL, M. J., JR. *Analyt. Chem.* 32 (1960) 1489

[17] KENDRICK, E. *Analyt. Chem.* 35 (1963) 2146

[18] REID, W. K., MEAD, W. L. and WEST, A. R. *Analyt. Chem.* 36 (1964) 1140

[19] REID, W. K. *Analyt. Chem.* 38 (1966) 445

[20] REID, W. K., MEAD, W. L. and WHYMAN, P. E. 'High Resolution Studies of High Molecular Weight Petroleum Fractions and Residues'. *Spectrosc. Grp Conf.* University College London (1965)

[21] TUNNICLIFF, D. D., WADSWORTH, P. A. and SCHISSLER, D. O. *Analyt. Chem.* 37 (1965) 543

[22] VAN KATWIJK, J. *Appl. Spectrosc.* 18 (1964) 102

[23] LUMPKIN, H. E. 'Trends in Physical Methods of Analysis of Petroleum and its Derivatives–Mass Spectrometry'. *7th Wld Petrol. Congr. Panel Discussion 15,* Mexico City (1967)

[24] POWELL, H. 'Integration and Co-ordination of Techniques'. *7th Wld Petrol. Congr. Panel Discussion 15,* Mexico City (1967)

[25] MERRITT, C., JR., ISSENBERG, P., BAZINET, M. L., MERRON, T. O., MURRAY, J. G. and GREEN, B. N. *Analyt. Chem.* 37 (1965) 1037

[26] McMURRAY, W. J., GREEN, B. N. and LIPSKY, S. R. *Analyt. Chem.* 38 (1966) 1194

[27] DORSEY, J. A., HUNT, R. H. and O'NEAL, M. J. *Analyt. Chem.* 35 (1963) 511

[28] ALTGELT, K. H. *J. appl. Polym. Sci.* 9 (1965) 3389

[29] HILLS, I. R. and WHITEHEAD, E. V. *Nature, Lond.* 209 (1966) 977

[29a] HILLS, I. R. and WHITEHEAD, E. V. 'Pentacyclic Triterpanes from Petroleum and their significance'. *3rd Int. Meet. Org. Geochem.* London (1966)

[30] BAKER, E. W. *J. Am. chem. Soc.* 88 (1966) 2311

[31] CLERC, R. J. and O'NEAL, M. J., JR. *Analyt. Chem.* 33 (1961) 380

[32] MEAD, W. L., WEST, A. R., HALLIDAY, J. S., RIDDOCK, A. and HOWARTH, H., 'A Combined Electron Impact–Field Ionization Source for Use with a High Resolution Mass Spectrometer'. *Am. Soc. Test. Mater. 15th A. Conf. Mass Spectrom. Allied Topics,* Denver (1967)

[33] BECKEY, H. D., KNOPPEL, H., METZINGER, G. and SCHULZE, P., *Advances in Mass Spectrometry,* Vol. 3, W. L. MEAD (Ed.), Institute of Petroleum, London, 1966

[34] WANLESS, G. G. and GLOCK, G. A., JR. *Analyt. Chem.* 39 (1967) 2

4.2

INORGANIC ANALYSIS BY SPARK SOURCE MASS SPECTROMETRY

R. Brown and D. Preston

4.2.1. INTRODUCTION

SPARK source mass spectrometry is well established as a highly sensitive method for the detection of constituent trace elements in many varied matrices. The inherent properties of approximate uniformity of sensitivity, complete coverage of the elements in one analysis, the linearity of response greater than 10^6 to one, facility of spectral interpretation, and the absence of prolonged presampling requirements, make the spark source mass spectrometer one of the most powerful tools available for general trace elemental analysis in solids, at this time.

In this paper, the development of the technique is traced, the current position is considered and results are given, to illustrate recent improvements of accuracy.

4.2.2. INSTRUMENTAL

The brilliant researches of Sir J. J. Thomson in 1913[1], into the behaviour of positive rays within applied magnetic fields, laid the foundations for the early instrumentation of Aston and Dempster, and the modern, commercial mass spectrometers.

Dempster, in 1935[2], indicated that the combination of the vacuum radio frequency spark source with a suitable mass spectrometer would be a sensitive analytical method of wide applicability. Hannay[3] in 1954, described a double focusing mass spectrometer, of Mattauch–Herzog geometry and obtained results that strongly endorsed these earlier predictions. Hannay and Ahearn[4] demonstrated that low detection limits of the order of 0·1 p.p.m. could be obtained with a mass spectrometer of this type, and that the detection sensitivity was approximately the same for all investigated elements.

The massive advances, achieved in semiconductor technology and solid state physics after 1945 revealed the necessity for new analytical techniques capable of detecting impurities at sub-p.p.m. level. The success of the above spark source mass spectrometric investigations prompted the advent of the first commercial, double focusing spark source mass spectrometer for inorganic trace elemental analysis.

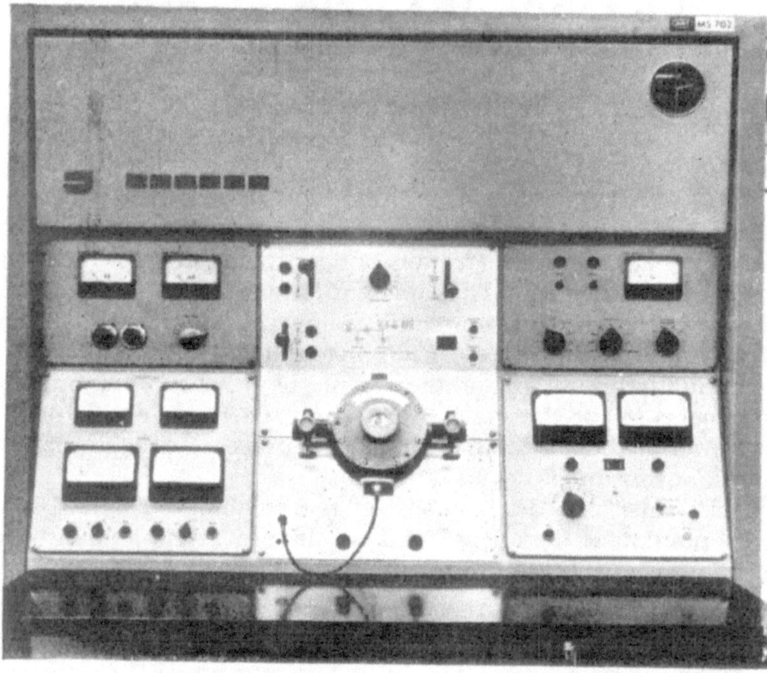

Figure 4.2.1. AEI MS702 *spark source mass spectrometer.* [*By courtesy of* GEC-AEI (Electronics) Ltd.]

The AEI MS702 spark source mass spectrometer, shown in *Figure 4.2.1*, is capable of detecting 0·001 p.p.m. This instrument is of near Mattauch–Herzog geometry, and is equipped with a pulsed r.f. ion source.

The ion path through the MS702 is shown in *Figure 4.2.2.*

Ions are produced by the pulsed radio frequency discharge of approximately 30 kV passed between two sample electrodes (0·5 × 0·45 in.) and are accelerated from the ionization region by an ion accelerating voltage of some 20 kV. The mode of ionization imparts a wide energy-spread to the ions and an electro-static analyser is necessary to bring the ions to focus. The ions are then mass analysed within the magnetic analyser and brought to a

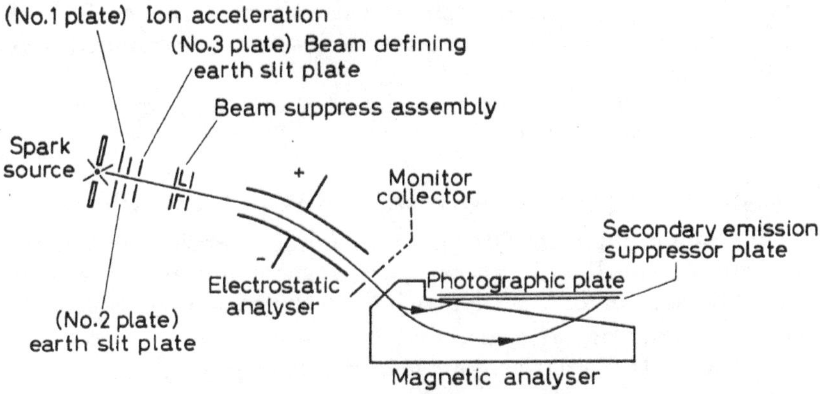

Figure 4.2.2. MS702 *schematic diagram.*
[*By courtesy of* GEC-AEI (Electronics) Ltd.]

planar focus. An Ilford Q2 photographic plate is placed within this focal plane and mass ranges of 40:1 ratio can be covered in one exposure (i.e. from lithium through uranium).

The photoplate position can be adjusted relative to the incident ion beam, and up to 30 graded exposures can be made on the same

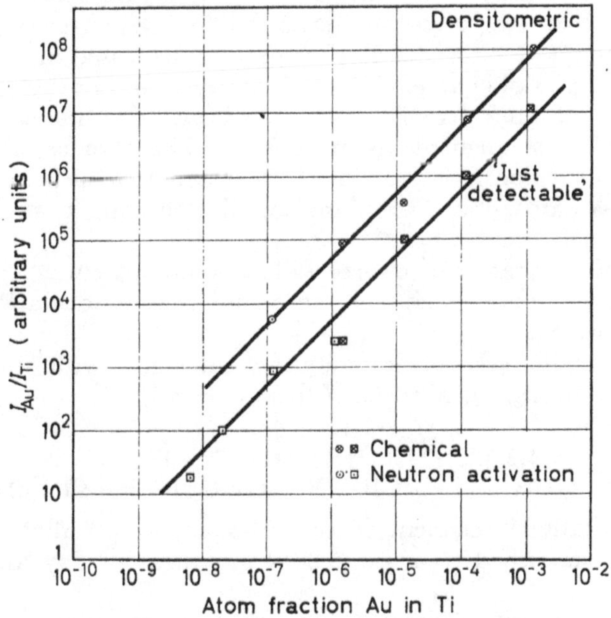

Figure 4.2.3. Linearity of response using Ilford Q2 plates.
[*By courtesy of* GEC-AEI (Electronics) Ltd.]

plate. The exposure length can be varied over a range of 10^7–1 and consequently, impurity concentrations can be determined over a similar range.

Figure 4.2.3 illustrates the linearity of the technique over such concentration ranges ($>10^5$:1).

4.2.3. IDENTIFICATION OF TRACE IMPURITIES

The first step in the identification of the isotopic species corresponding to the various lines in an exposure, is the calibration of a mass scale. The position of a particular line on the photoplate is proportional to $K\sqrt{m/e}$ or, to a reasonable approximation, m/e for the relevant ion. It is relatively easy to identify two or three lines in the spectrum and subsequent mass scale evaluation can be achieved. It is normal operational practice to maintain constant magnetic field conditions so the comparison of similar plates is a simple matter.

Figure 4.2.4 illustrates a typical mass spectrum of stainless iron, covering a mass range of 7–280. The strongest line occurs at mass 56, the major iron isotope, and the intensity pattern of the lines at masses 54, 56, 57 and 58 correspond to the abundances of the various iron isotopes. At masses 27, 28, 28½ and 29 a similar pattern is observed, corresponding to the doubly-ionized iron. The phenomenon is repeated for Fe^{3+} and, with decreased intensity, for higher multiples of charge. Other obvious features of this plate are the groups of lines for tin (Sn) and Lead (Pb) indicating the presence of these elements as impurities. The absence of lines in the 140–150 region is indicative of the absence of such rare earth elements as Samarium, Europium and Gadolinium, from the iron matrix.

The series of graded exposures will show up any surface contamination and inhomogeneity of the samples when deviation from linearity occurs.

A regular increase of intensity for all lines, with increasing exposure, confirms the absence of these effects.

4.2.4. THE DETERMINATION OF TRACE IMPURITY CONCENTRATIONS

The quantitative assessment of trace impurities, once their identity has been established, is a natural requirement of any analytical technique.

The impurity concentrations may be estimated by the visual method, a rapid semi-quantitative method, or by using a recording

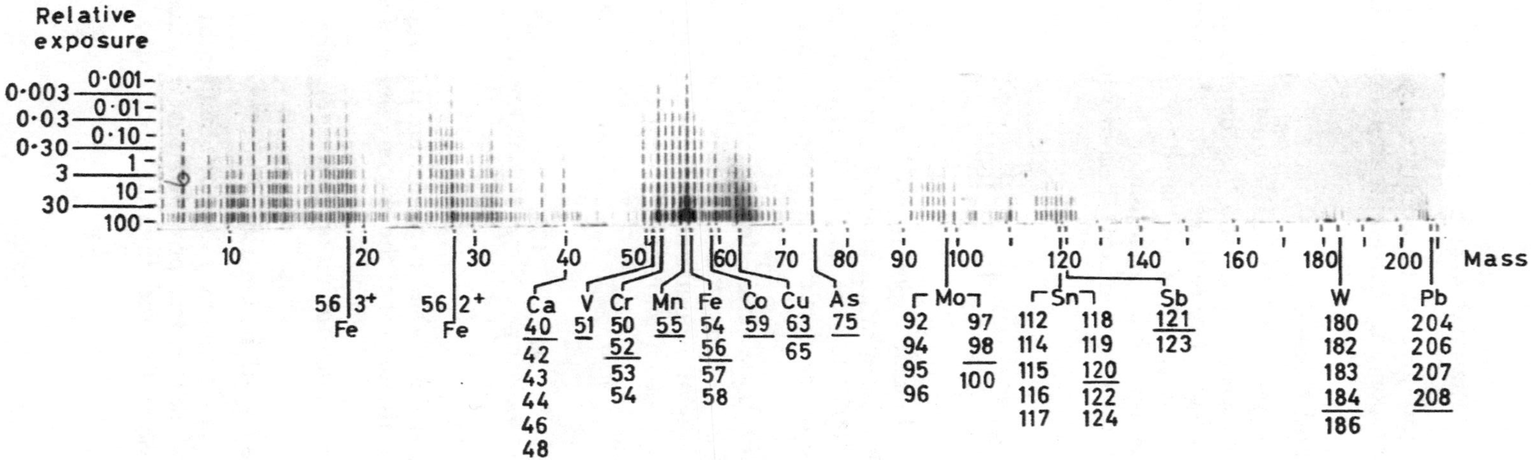

Figure 4.2.4. Mass spectrum of stainless iron.
[*By courtesy of* AEI-GEC (Electronics) Ltd.]

microdensitometer to give more accurate measurement of line density with a corresponding increase in analytical precision.

4.2.4.1. The Visual Method

The method is based on the visual estimation of the exposure required to produce selected lines at a 'just detectable' density and accuracies within a factor two to three can be obtained in this way. The photoplate sensitivity (S_p) is evaluated by noting the exposure (E_s) at which the line, due to a minor isotope of the major constituent (i.e. an internal standard), becomes 'just detectable'.

$$S_p = \frac{E_s}{E_{max.}} \times \frac{X}{100} \times \frac{I_s}{100} \times 10^6 \qquad \ldots (4.2.1)$$

where $E_{max.}$ is the longest exposure on the photoplate, X is the atomic percentage concentration of the internal standard element and is normally 100% for the major constituent, and I_s is the isotope abundance of the particular isotope chosen as the internal standard.

The value of the plate sensitivity is obtained in terms of the actual concentration (p.p.m. atomic) of any isotope, which is just detectable on the longest exposure on the photoplate.

The concentration of an impurity element C_i can be calculated from the equation:

$$C_i = S_p \times \frac{E_{max.}}{E_i} \times \frac{100}{I_i} \text{ p.p.m. atomic} \qquad \ldots (4.2.2)$$

where E_i is the 'just detectable' exposure for the impurity isotope and I_i is the isotope abundance of the chosen isotope of the impurity element.

Equation 4.2.2 is applied to first order spectra only.

4.2.4.2. Analysis by Microdensitometer

Improved quantitative accuracy may be obtained by the application of the microdensitometer, provided that relative sensitivity values are established for the internal standard and impurity elements.

Impurity concentrations are calculated as follows:

$$C_i = \frac{E_s}{E_i} \times \frac{X}{100} \times \frac{I_s}{I_i} \times \frac{S_s}{S_i} \times \frac{A_i}{A_s} \times \frac{M_i}{M_s} \times 10^6 \text{ p.p.m. atomic} \quad \ldots (4.2.3)$$

where E_s and E_i are the exposures at which the lines for the internal standard and the isotope of the impurity element have identical density values, X = atomic percentage concentration of the internal standard element (normally the major component isotope).

I_s and I_i are the isotopic abundances of the internal standard and impurity isotopes. S_s and S_i are the relative sensitivities of the singly-charged ions of the internal standard and impurity elements, A_s and A_i are the relative areas of the mass spectral lines of the standard and impurity elements and M_s and M_i are the ratios of the intensities of singly-charged ions to those of the particularly multiply-charged ions on which the estimate is based. If first order lines are taken then this last ratio is unity.

4.2.5. ACCURACY AND REPRODUCIBILITY

The early work of Gorman *et al.*[5], utilizing electrical detection to measure impurity concentrations in steel, illustrated reproducibility values of the order of $\pm$ 3%. Craig *et al.*[6] recorded precision levels of $\pm$ 20% using photoplate detection and the measurement of close isotope ratios has revealed reproducibility of measurement to be $\pm$ 3–4%. Subsequent investigations have been directed, therefore, into the instrumental contribution to error, and also into the problem of sample heterogeneity.

The work of Halliday *et al.*[7], thoroughly examined the effect of variation of the source parameters upon standard deviation values. A basic limit of attainable reproducibility was obtained by repeating 15 exposures across a photoplate with identical sparking conditions. The standard deviation on the ratio of $^{107}Ag^+/^{109}Ag^+$, present in Johnson and Matthey copper standard CA2, was 3·7%. The same exposure on eight different plates gave a value of 5·8% showing the variations from plate to plate.

The accelerating voltage, spark pulse length and repetition rate were maintained constant for a series of experiments with aluminium, copper and steel standards. The standard deviations so obtained, varied between 9% and 21%. The average standard deviation for repeat exposures was 15 $\pm$ 6%. When the pulse repetition rate and pulse length were varied the average value was 30 $\pm$ 5%.

The consumption of sample is of the order of a few milligrammes during the sparking process. Consequently, to maintain good reproducibility, a representative portion of the sample must be taken, and thus the homogeneity requirement is high. This is particularly relevant when the impurity concentration is large and segregation is, unfortunately, most probable.

An AEI SEM2 microanalyser was used to examine various standards and a resulting correllation was drawn between poor reproducibility and sample inhomogeneity. For example, in aluminium sample AC1, titanium aggregation was detected and measured by the X-ray microprobe analyser and, from theoretical considerations, the relative intensities of the $^{48}Ti^{+}$ spectrum were expected to lead to 12–1 variations. Subsequent mass spectrometric measurement of these line intensities, verified these expectations.

Several workers[8] have demonstrated the applicability of spark source mass spectrometry to geological problems of trace element analysis. The sensitivity of the technique has proved particularly useful, but in order to make high quality geochemical investigations, a precision somewhat greater than 20% was required. Nicholls *et al.*[9] describe a fusion technique whereby the problem of sample inhomogeneity has largely been overcome, and report standard deviations of $\pm$ 5%.

The sample is mixed with a flux of known content, fused at 900°C within a platinum foil capsule, cooled quickly and the resulting glass ground in an agate mortar. The process of fusion and grinding is repeated several times to ensure complete mixing of the sample material and then the powdered fusion material is mixed with graphite and suitable electrodes are prepared for the mass spectrometer. The internal standard is rhenium, and is added in such quantities that the final mix contains approximately 20 p.p.m. of rhenium. Results were obtained for 35 element impurities in rocks and precisions of less than 5% maintained, on impurity concentrations in the range 0·7 p.p.m. to 200 p.p.m. weight.

The above work has amply demonstrated the effect of inhomogeneity of trace impurity distribution upon the precision of spark source mass spectrometry as an analytical technique. However, the elaborate sample preparation techniques described above are not applicable to metallurgical investigations, and a more general method of overcoming the problem of inhomogeneity has been sought, that could achieve precision levels comparable to those reported by Nicholls, but precluding the lengthy sample preparation technique.

4.2.5.1. Factors affecting the precision of the method

It was considered that the following factors contributed to the large standard deviation of 15% for 27 elements in titanium dioxide analyses.

(*i*) The homogeneity of the sample.

(*ii*) The inability to measure accurately the charge laid down on the photoplate during short exposures.

(*iii*) The variation of the spark pulse length and repetition rate.

(*iv*) The inherent errors of photoplates.

Hence, a method of photoplate exposure, where source conditions were maintained constant and that also consumed more sample in the shorter exposures, as well as the long ones, was required.

The 'ion beam chopper', a recent instrumental development which enables these requirements to be fulfilled simply, both in principle and operation, has been reported by Halliday *et al.*[10]

The normal method of exposure allowed the ion beam to fall continuously on the photoplate until the preselected charge had been recorded. The exposure was terminated by the application of +1300 V between a pair of ion beam suppress electrodes, situated behind the source in the MS702 mass spectrometer.

In the modified system, a pulsed voltage of 200 V is applied to the beam suppress plates. The width of the pulse is fixed at 5 μs while the frequency is variable from 0·5–200 kHz.

The pulses are fed from a commercial pulse generator, with variable frequency and pulse width, amplified by a transistorized

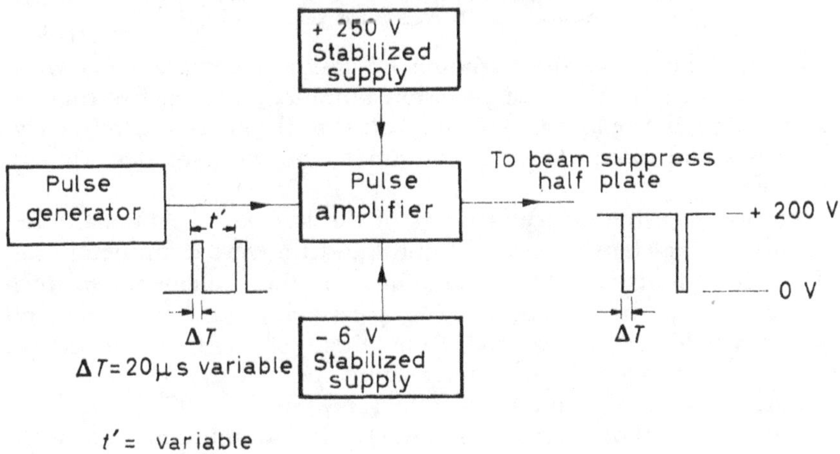

Figure 4.2.5. Schematic diagram of pulse voltage supply circuits.
[*By courtesy of* GEC-AEI (Electronics) Ltd.]

amplifier which produces +200 V, and these pulses are supplied to the beam suppress assembly.

Figure 4.2.5 outlines the schematic layout of the necessary supplies to provide the pulsed voltage.

Jackson, Vossen and Whitehead[11], have reported standard deviation improvements from ±15% to ±5% on a series of titanium dioxide powder analyses with the ion beam chopper modification fitted to the mass spectrometer.

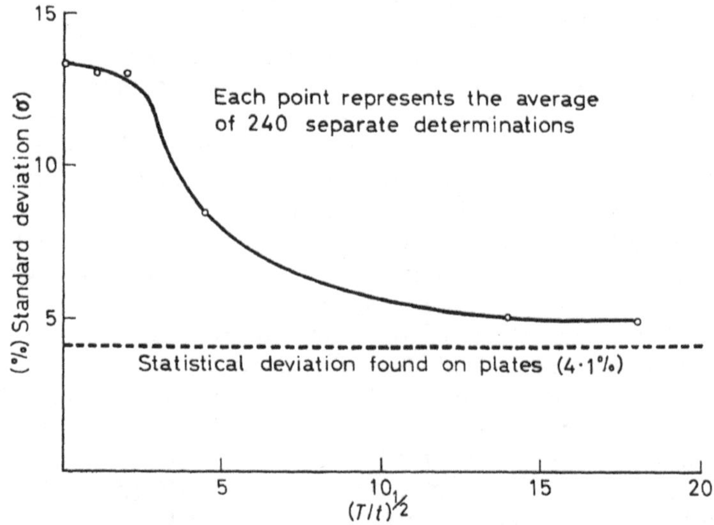

Figure 4.2.6. Variation of standard deviation with $(T/t)^{\frac{1}{2}}$.
[*By courtesy of* GEC-AEI (Electronics) Ltd.]

Figure 4.2.6 shows the variation of standard deviation (σ) with $(T/t)^{\frac{1}{2}}$ and shows that the precision obtained, approaches that of the limiting photoplate error of 4–5% (i.e. as the chopping frequency decreases and more sample is consumed, the standard deviation is improved).

To optimize the critical settings of the ion beam chopper pulse frequency, the elements lead, bismuth and tin were examined in an aluminium matrix. The insolubility of these elements in this particular matrix has been well accepted, and inhomogeneity effects would be anticipated from these elements. Precisions, within the range 25–30% have been previously reported[7] for impurities in metal and non-metallic samples.

Table 4.2.1 shows results obtained with aluminium and with the ion beam chopper facility. An overall improvement in precision is noted.

Table 4.2.1. Percentage Standard Deviation for Aluminium Standard 1791

Element	52 Cr internal standard			
	Plates 1–8 8 *plates in* 1 *day*	*Plate* 9, 11, 13–23 8 *plates*, 1 *daily*	*Plate* 10, 12, 14–24 8 *plates*, 1 *daily*	*Plates* 9–10, 11–12, *etc.* 1 *pair daily*
^{48}Ti	4·1	6·2	7·7	5·6
^{51}V	6·2	9·4	10·8	6·1
^{55}Mn	7·4	4·5	7·1	4·0
^{56}Fe	10·3	8·7	7·7	6·1
^{58}Ni	9·6	9·6	9·1	7·7
^{63}Cu	11·9	12·0	9·0	6·3
^{64}Zn	12·0	7·1	8·6	6·9
^{68}Ga	7·1	7·2	9·8	4·1
^{120}Sn	16·6	19·3	20·1	17·7
^{208}Pb	19·7	16·1	18·2	4·8
^{209}Bi	20·2	15·1	17·4	17·2
Average	11·4	10·5	11·4	7·9

Figure 4.2.7 shows plots of standard deviation obtained against pulse frequency for these elements. The lowest standard deviations

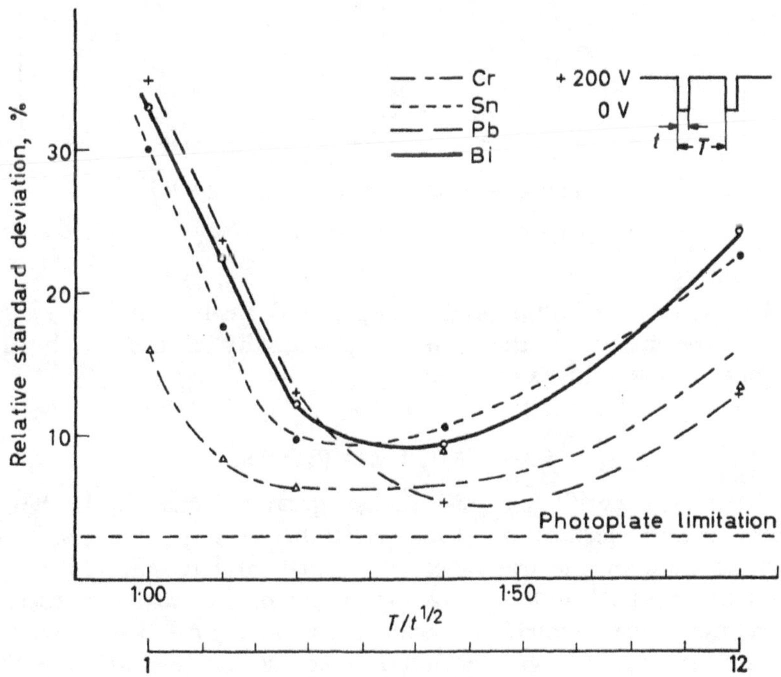

Figure 4.2.7. Variation of percentage standard deviation with chopping frequency

are obtained when an ion beam chopper pulse frequency of 100 kHz is used.

From a comparison of the two gamma curves in *Figure 4.2.8* two improvements are noted. A more accurate assessment of the slope of the curve can be obtained and thus isotope intensity ratios can be estimated more accurately and the concentration range, over which these accurate measurements can be achieved, is extended.

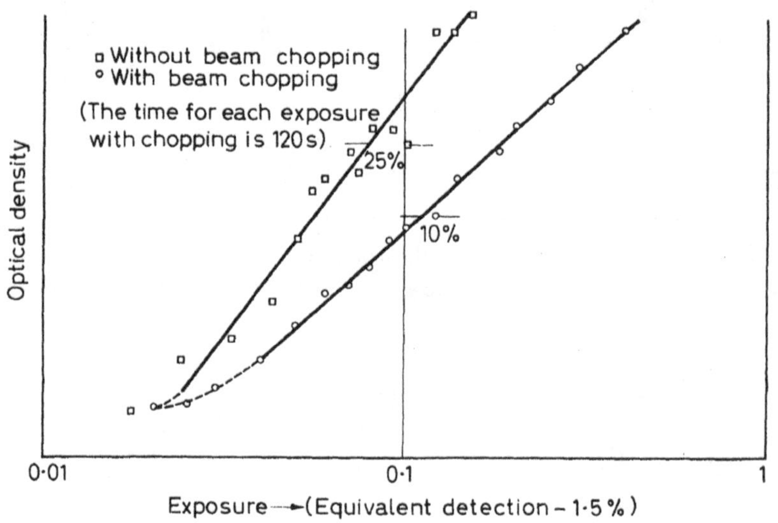

Figure 4.2.8. Single gamma curves of one isotope

The analyses of aluminium, copper and steel have borne out early expectations of the general applicability of the ion beam chopper to the analysis of metals.

4.2.6. APPLICATIONS

As general scientific progress makes greater demands, in both tolerance and specification of materials, there is a growing requirement to monitor the trace element content and its effect, if any, upon the physical and chemical properties of the parent matrices. Inorganic trace element occurrence, in biological and natural systems, is assuming greater importance, as biochemical investigations reveal its relation to dynamic and living processes. Consequently, a technique that found its early application in metallurgical and semiconductor fields is now being used in many different spheres of interest. Some examples are given below.

4.2.6.1. Metallurgy

The analysis of most common metallurgical matrices has been well documented. Spark source instruments are being used to examine copper, aluminium, steel, and nickel, for example.

4.2.6.2. Semiconductor Fields

The electrical properties of semiconducting materials are closely related to the trace impurity content. Spark source mass spectrometry has been used to monitor trace impurities in such materials as gallium arsenide, germanium silicide and indium arsenide.

4.2.6.3. Pigment Analysis

Jackson and Whitehead[12] have established a routine analytical technique for titanium dioxide pigments. Quantitative data on 27 elements can be obtained in 3 h.

4.2.6.4. Geology

The successful analysis of rocks and ores has been discussed earlier in the text. Important geochemical interpolations as to the presence of valuable ore bodies, and petroleum deposits, can be drawn from the trace element pattern in rocks.

4.2.6.5. General Inorganic Chemical Investigations

Some recent analyses include thorium oxide, europium oxide, holmium, europium metal and yttrium oxide.

4.2.6.6. Biological Samples

These include blood plasma samples, human hair, cow tooth plaque, and elements in pharmaceutical products.

4.2.6.7. Liquids

By a small modification to the source of the MS702, liquids can be investigated. The principle involves the deposition of a liquid onto a gold disk which, after evaporation of the fluid, is mounted in the source and then sparked against a counter electrode. Stream waters have been analysed in this manner and sensitivities as high as 5×10^{-11} g/ml have been obtained.

4.2.6.8. Small Samples

A tipped-electrode technique has been developed to handle limited quantities of sample material. The lowest limit of detection from a 1 mg sample is 0·03 p.p.m. atomic, although limits lower than this should be readily attainable.

4.2.7. SUMMARY

To summarize, this analytical technique of unique sensitivity, linearity and variety of application, is now established on a fully quantitative basis. The continual development of sample handling techniques has widened the scope of application and it is to be expected that new opportunities will be revealed in hitherto unexplored fields, as analysts exploit spark source mass spectrometry in accurate elemental analysis.

The writers thank AEI Electronics Group for permission to publish this paper.

REFERENCES

[1] THOMSON, J. J., *Rays of Positive Electricity*, Longmans Green, New York and London, 1913
[2] DEMPSTER, A. J. *Proc. Am. phil. Soc.* 75 (1935) 755
[3] HANNAY, N. B. *Rev. scient. Instrum.* 25 (1954) 644
[4] HANNAY, N. B. and AHEARN, A. J. *Analyt. Chem.* 26 (1954) 1056
[5] GORMAN, J. G., JONES, E. J. and HIPPLE, J. A. *Analyt. Chem.* 23 (1951) 438
[6] BROWN, R., CRAIG, R. D. and ELLIOTT, R. M., *Advances in Mass Spectrometry*, Vol. 2, Pergamon, Oxford, 1963
[7] HALLIDAY, J. S., SWIFT, P. and WOLSTENHOLME, W. A., *Advances in Mass Spectrometry*, Vol. 3, W. L. MEAD (Ed.), Institute of Petroleum, London, 1966
[8] TAYLOR, S. R. *Geochim. cosmochim. Acta* 29 (1965) 1243
[9] NICHOLS, G. D., GRAHAM, A. L., WILLIAMS, E. and WOOD, M. *Analyt. Chem.* 39 (1967) 584
[10] BINGHAM, R. A., BROWN, R., HALLIDAY, J. S., POWERS, P. and VOSSEN, P. G. T. 'Improvements in Quantitative Impurity Analysis of Solids by Mass Spectrometry'. *International Conference on Characterization of Materials*, Pennsylvania State University (1966)
[11] JACKSON, P. F. S., WHITEHEAD, J. and VOSSEN, P. G. T. In press
[12] JACKSON, P. F. S. and WHITEHEAD, J. *Analyst, Lond.* 91 (1966) 418

DISCUSSION

R. J. HERITAGE (R.R.E.)

What do you think is the mechanism for the improvement in precision by using the chopping technique?

Reply. The technique uses more sample, due to the longer exposure time, and hence longer sparking time is necessary. Conditions are thus similar to emission spectroscopy, where excitation conditions are maintained constant.

4.3

AN EXAMINATION OF METAL CHELATES BY MASS SPECTROMETRY

A. E. Jenkins and J. R. Majer

4.3.1. INTRODUCTION

It is many years since the volatility of a number of metal chelates was first demonstrated by Morgan but it is only comparatively recently that this property has been exploited for the detection and estimation of metals. The most striking example of this exploitation is the separation by gas phase chromatography of a number of the substituted acetyl acetonates of different metals[1]. A variety of metal chelates is sufficiently volatile and stable to be admitted to the source of a mass spectrometer, although it may be impossible for them to pass unchanged through a gas chromatography column.

The mass spectrum of a metal chelate, in addition to providing a simple means of identification, can give information about the structure of the molecule itself. The structures of many metal chelates are still a matter of debate, but in attempting to use the mass spectrometer for structure determination, it is necessary to correlate the mass spectrum obtained from molecular species in the gas phase, with those molecular species existing in the solid phase before evaporation. Many metal chelates are hydrated and lose water at comparatively low temperatures. It is likely, therefore, that this process of dehydration will precede evaporation so that a mass spectrum containing a hydrated metal chelate ion peak would not be anticipated. Further, because the water molecules would be comparatively weakly bonded to the metal chelate, it is unlikely that even if the hydrated metal chelate could survive evaporation, the molecular ion would be sufficiently stable to make a contribution to the mass spectrum. In fact, using a conventional mass spectrometer and operating the source at 70 V, we have not observed any hydrated metal chelate ions. Simple dehydration is not a serious complication, but there is also the possibility of hydrolysis and thermal decomposition, preceding evaporation.

When there is a satisfactory correlation between the mass spectrum and the composition of the solid phase it is possible to consider estimating the amount of metal chelate evaporated and hence the amount present by a measurement of the ion current produced. It has been common practice, when carrying out quantitative analysis with a mass spectrometer, to provide a reservoir of sample vapour, the pressure of which remains substantially constant while a spectrum is recorded. The partial pressure of a component in the sample is then proportional to the height, on the recorder trace, of mass peaks, which are due to that component only. The proportionality constant or sensitivity for each peak is determined by recording separately the mass spectrum of the pure component at a known pressure. The only satisfactory method of recording mass spectra of metal chelates is by their direct evaporation into the ion source from a probe which can be inserted into the vacuum system of the mass spectrometer.

When such a device is used, however, two problems arise. The partial pressure of the metal chelate changes during the time required to record a mass spectrum, so that it is no longer possible to apply previously determined sensitivities and, in addition, at very low concentrations of the metal chelates, evaporation may be complete before the mass spectrum has been recorded. There is a further instrumental complication, in that the sensitivity of the instrument may change owing to deliberate or unintentional variations in the gain of the electron multiplier. It is necessary, therefore, to provide an internal standard of which the partial pressure is known and remains constant, and of which the mass spectrum may be recorded at the same time as that of the metal chelate. This may be most simply achieved by providing a reservoir of the vapour of heptacosafluorotri-*n*-butylamine (a compound commonly used as a mass marker in mass spectrometry) at a known pressure and allowing this to leak into the ion source at a constant rate at the same time as the metal chelate is evaporating.

As the complete mass spectrum of the metal chelate at low concentration cannot be recorded in the evaporation time, and because, even if this were possible, the mass peak sensitivities would not remain constant, one must use a single mass peak characteristic of the metal chelate and record the rise and fall of the ion current at this particular m/e value during the course of the evaporation. The integrated ion current would then be directly proportional to the amount of the metal chelate evaporated, and hence to the concentration of metal in the original sample, provided that no irreproducible thermal breakdown occurs during evaporation. The most suitable ion for this purpose would be that most characteristic of the metal chelate, and hence the molecular ion itself would be most desirable.

There is an instrumental difficulty, associated with such a recording, in that no magnetic scanning spectrometer can provide a means of tuning-in the ion current at a selected *m/e* value before the ion current itself has been established. Thus, a method had to be found in which the instrument could record the ion current at a significant *m/e* value before, during and after evaporation.

The method employed was to use a second substance, in this case heptacosafluorotri-*n*-butylamine, which did not exhibit a spectrum containing a peak at this *m/e* value. A constant partial pressure of this substance is maintained within the ion source, using the conventional gas handling facilities of the instrument. It is then possible to select a peak, from the mass spectrum of this compound, which is close to the significant *m/e* value. The peak selected is then displayed on the oscilloscope facilities of the instrument by the normal tuning-in procedure. Using the peak comparison technique, it is now possible to display alternately on the oscilloscope, the reference peak and a sample peak. This is achieved on the model MS9 by adjusting the decadic resistance to a value given by the ratio of the precise masses of the ions responsible for the peaks.

A recorder trace taken at this stage shows only the reference peak repeated at intervals. When the sample is lowered into the heating region of the ion source and evaporation begins, a second peak appears between the repeated reference peaks. In sharp distinction from the latter, which has a constant height, this second peak increases in height, reaches a maximum and then decays and finally disappears. It is the envelope of these secondary peaks which is a record of the rise and fall of the peak being measured and it is therefore the area under this envelope which is proportional to the integrated ion current and hence to the concentration of metal chelates.

It can now be seen that the reference compound serves a two-fold purpose: it enables the mass scale to be calibrated so that the rise and fall of the selected sample peak can be completely recorded and it provides an internal standard against which any variation of instrument sensitivity may be measured. *Figure 4.3.1* shows such a record for nickel dimethylglyoximate.

The alternative procedure of recording continuously the peak height at the significant *m/e* value was discarded because instrumental drift caused peak height to vary during evaporation. The peak comparison technique, although providing an interrupted rather than a continuous trace, ensured that the full peak height was always recorded.

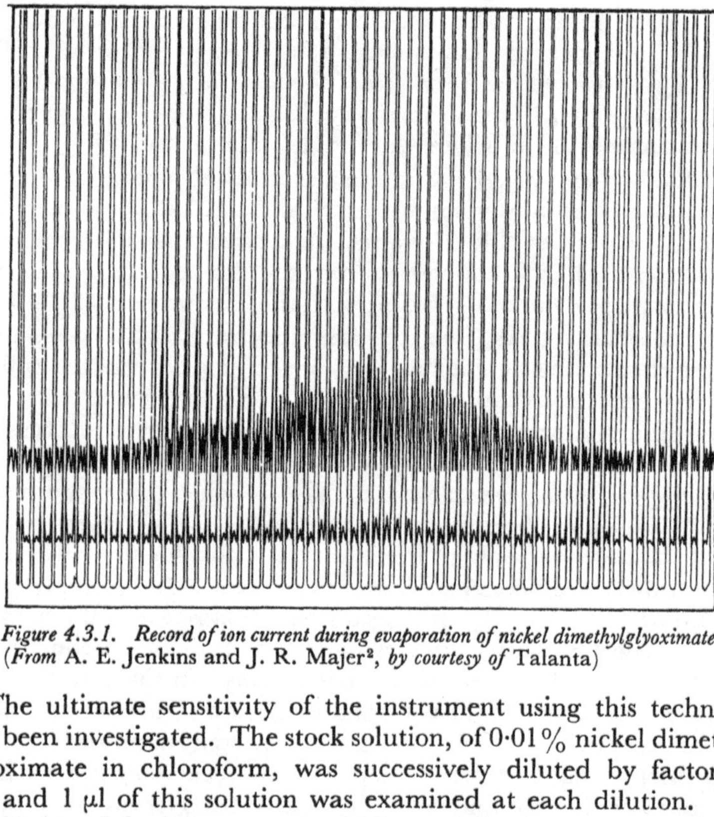

Figure 4.3.1. Record of ion current during evaporation of nickel dimethylglyoximate. (*From* A. E. Jenkins and J. R. Majer[2], *by courtesy of* Talanta)

The ultimate sensitivity of the instrument using this technique has been investigated. The stock solution, of 0·01% nickel dimethylglyoximate in chloroform, was successively diluted by factors of 10, and 1 μl of this solution was examined at each dilution. The sensitivity of the instrument could be varied, over several orders of magnitude, by adjustment of the electron multiplier gain and the electron beam current. As the gain is increased, the signal-to-noise ratio is reduced, but at the highest usable gain the sample size could be reduced to 10^{-14} g.

With the electron multiplier gain constant, a series of determinations was made for different sample sizes. When the area under the envelope is plotted against sample size a straight-line relationship is obtained.

It is the purpose of this paper to examine the correlation of mass spectra of the oxinates of a number of metals with a composition of the solid metal chelate and to attempt to apply the integrated ion current method, outlined above, as first described by Jenkins and Majer[2], to the determination of submicrogramme quantities of metal oxinates.

4.3.2. EXPERIMENTAL

All measurements were made with an AEI MS9 mass spectrometer, fitted with an insertion lock and a direct evaporation probe. Samples were admitted in pyrex glass tubes, 27 mm long and 1 mm internal diameter, which were fitted to the end of the probe and discarded after use. The tubes were injected with 5 μl of the metal chelate solution by means of a Hamilton syringe (701–*NW/G*), alternatively the chelate was precipitated directly by mixing 1 μl of an aqueous solution of the metal salt with 1 μl of an alkaline solution of the chelating agent in the tube.

4.3.2.1. Reagents

All metal oxinates were prepared by methods described elsewhere[3]. Solutions were made up in chloroform. Insoluble oxinates were precipitated directly from aqueous solution in the probe. Heptacosafluorotri-*n*-butylamine was supplied by Koch–Light Laboratories Ltd.

4.3.2.2. Procedure

A measured volume (1–5 μl) of the metal chelate solution was injected into the sample tube, which was then transferred to the vacuum system of the mass spectrometer, through the insertion lock, and held in the cool part of the system. A measured pressure of heptacosafluorotri-*n*-butylamine was admitted to the gas inlet system. A reference peak was selected and displayed on the oscilloscope. The decadic resistance was then set to an appropriate value and the peak switching system so adjusted that the oscilloscope gave an alternate display of the reference and sample peaks. At this stage the sample peak should not appear. The galvanometer recorder was set in operation, so as to record only the peak pair displayed on the oscilloscope. The sample tube was then lowered into the heated region of the source (approximately 250°C) to allow evaporation of the metal chelate.

When evaporation begins, the sample peak appears on the oscilloscope and is recorded. Evaporation is complete in 10–20 s depending on sample size, and the sample peak then disappears from the oscilloscope and the recorder trace. The amount of metal chelate evaporated is proportional to the integrated ion current at the significant *m/e* value, and therefore to the area under the envelope of the sample peaks on the recorder trace. When the complete mass spectrum was to be recorded and the evaporation temperature determined, the sample was lowered into the heating zone and the temperature raised until evaporation commenced. The position of the probe was then adjusted to keep the total ion current constant while the mass spectrum was recorded.

4.3.3. RESULTS

4.3.3.1. Copper Oxinate

The structure is given in the literature as $CuOx_2 2H_2O$. This begins to dehydrate at 60°C, and above 100°C exists as $CuOx_2$. At temperatures above 300°C, the complex decomposes, forming copper oxide at 700°C. Evaporation in the mass spectrometer source was rapid at 255°C and the spectrum obtained consistent with the structure $CuOx_2$, with peaks appearing at *m/e* 351 ($^{63}CuOx_2$), 207 ($^{63}CuOx$) and 63 (Cu). The integrated ion current at *m/e* 351 was recorded for a range of sample sizes and the results were plotted as log peak area against log sample weight (*Figure 4.3.2*). A straight-line relationship is obtained and the reproducibility at the 10^{-10} g level is $\pm 20\%$. The sensitivity may be

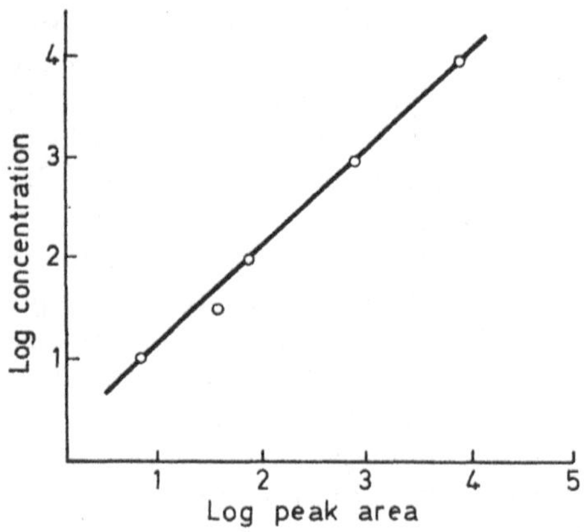

Figure 4.3.2. Relationship between integrated ion current (m/e = 351) and sample size for copper oxinate

most conveniently expressed as $x/y \times 10^{-12}$ where x is the peak area on the recorder chart, in square millimetres and y is the sample weight in grammes. The sensitivity (s) for copper oxinate, at *m/e* 351 and low gain, was 0·057.

4.3.3.2. Aluminium Oxinate

The complex becomes anhydrous above 100°C, but does not begin to decompose until a temperature of 375°C is reached. Decomposition is slow and incomplete until 1000°C. Evaporation in the source begins at 400°C, and the mass spectrum is in accordance with the structure $AlOx_3$ with peaks appearing at m/e 459 ($AlOx_3$), 315 ($AlOx_2$) and 171 (AlOx). The sensitivity at m/e 459 (s_{459}) is 0·01 and the limit of detection on low gain is 5×10^{-10} g.

4.3.3.3. Zinc Oxinate

Zinc oxinate is said to exist as the dihydrate at room temperature and as $ZnOx_2$ above 135°C. It begins to decompose above 300°C, giving the oxide at 700°C. Evaporation in the source does not begin until 380°C and the mass spectrum is no longer simple, as in the previous two cases. In addition to peaks corresponding to the structure $ZnOx_2$ at m/e 352 and ZnOx at 208, there is a peak at m/e 416 corresponding to the ion Zn_2Ox_2. The exact structure of this ion is not easy to formulate, but the zinc atoms must bridge the two oxine residues to which they are only singly bonded. Since evaporation is taking place at a temperature well above the decomposition point, the ion must be formed from molecules or radicals produced in the decomposition process. The decomposition is reflected in the very low sensitivity obtained ($s_{352} = 3{\cdot}5 \times 10^{-4}$). The limit of detection is of the order of 10^{-8} g.

4.3.3.4. Magnesium Oxinate

The anomalies noted for zinc oxinate are even more marked in the case of magnesium oxinate. This exists as the tetrahydrate at room temperature, as the dihydrate above 100°C and in the anhydrous form above 155°C. Between 160°C and 345°C it undergoes decomposition. Evaporation in the source did not take place until a temperature of 400°C was reached and the mass spectrum exhibited peaks at m/e 480 due to the ion $Mg_2Ox_3^+$ and at 346 due to the ion $Mg_2Ox_2^+$. Thus it seems that the decomposition occurs by a polymerization process in which oxine is lost. This process is even more marked in the following examples.

4.3.3.5. Neodymium Oxinate

There has been some difference of opinion about the existence of neodymium oxinate of constant composition, although samples corresponding to the formula $NdOx_3$ have been prepared. The complex was not considered to be stable and so the mass spectrum is of particular interest. Evaporation into the source occurred at a temperature of 400°C, and the mass spectrum was very complex. Among the mass peaks which could be identified were those at *m/e* 286 (NdOx), 430 ($NdOx_2$), 574 ($NdOx_3$) and 716 (Nd_2Ox_3), but there are also peaks between 800 and 1050 which could not be positively identified without mass measurement but are probably due to Nd_3Ox_3, Nd_3Ox_4 and Nd_4Ox_4. It is not possible to say, with certainty whether these peaks are due to thermal polymerization at a temperature of 400°C or whether the oxinate prepared contains polymer.

A similar spectrum is obtained with dysprosium oxinate and yttrium oxinate, the latter exhibiting a spectrum extending to *m/e* 1500.

4.3.3.6. Transition Metals

The oxinates of iron, cobalt, nickel and manganese all evaporate into the ion source at temperatures above the decomposition temperatures quoted in the literature and, in each case, the mass spectrum provides evidence of polymerization with the appearance of mass peaks corresponding to the ions $M_2Ox_2^+$. The sensitivities for the transition metals are all very low and of the same order (^{59}Co–$s_{347} = 2{\cdot}0 \times 10^{-4}$; ^{58}Ni–$s_{346} = 3{\cdot}0 \times 10^{-4}$; ^{56}Fe–$s_{344} = 2{\cdot}0 \times 10^{-4}$; ^{55}Mn–$s_{343} = 7{\cdot}0 \times 10^{-4}$).

4.3.3.7. Beryllium Oxinate

There has been some considerable discussion about the structure of beryllium oxinate and the most recent study, using the techniques of differential thermal analysis and infra-red absorption, suggests the formula $Be_2Ox_22H_2O$. At temperatures between 180–270°C one molecule of water is lost and at higher temperatures, hydrolysis and loss of oxine is said to take place with the final formation of beryllium oxide. Evaporation into the source took place at a temperature of 380°C and the mass spectrum showed intense peaks at *m/e* 297 ($BeOx_2$) and 153 (BeOx). This provides no support for the suggested structure[4] and is at variance with the proposed mode of decomposition. There is, however, evidence of polymerization provided by the appearance of a peak at *m/e* 450 corresponding to the ion $Be_2Ox_3^+$. The sensitivity is also very low ($s_{297} = 4{\cdot}0 \times 10^{-4}$).

4.3.4. Summary

The sensitivities mentioned above vary widely and are obviously affected by decomposition processes occurring during evaporation. It is possible to decide upon the suitability of a particular chelating agent, for the determination of the metal using this technique, by comparing the sensitivity with that of a standard substance. We have chosen nickel dimethylglyoximate as a standard metal chelate because of its high volatility, simple mass spectrum and high sensitivity. On this basis only the copper and aluminium oxinates have comparable low gain sensitivities. Use of high instrumental gain increases sensitivities one-thousandfold.

The authors are indebted to Dr. W. I. Stephen for supplying some of the oxinates and to Miss M. Reade for carrying out integrated ion current measurements.

REFERENCES

[1] MOSHIER, R. W. and SIEVERS, R. E., *Gas Chromatography of Metal Chelates*, Pergamon, Oxford, 1965
[2] JENKINS, A. and MAJER, J. R. *Talanta* 14 (1967) 777
[3] HOLLINGSHEAD, R. G. W., *Oxine and its Derivatives*, Butterworths, London, 1956
[4] SASTRI, M. N. and PRASAD, T. P. *Talanta* 14 (1967) 481

4.4

DATA HANDLING AND INSTRUMENTATION WITH THE A.W.R.E. MASS SPECTROMETERS

W. A. P. Young and B. W. J. Gooding

OUR data handling[1] for mass spectrometry at Aldermaston runs from measuring a chart with a ruler to punching paper tape automatically and sending it straight off to the computer (Atlas). If I may start with the ruler: one of our AEI MS5 machines is still sometimes used this way, chiefly as a check on more sophisticated methods. We have a form on which the measured peak heights can be entered and this is sent off to the computer. This may seem more difficult than just working it out on a hand machine, but we do get the benefit of error computation (we use our programme ISOTOP) and the standard output format which can go straight to the customer without further hand writing. Next, I should like to mention our two, two-stage machines which use peak switching by varying the accelerating voltage, and pulse counting either from an electron multiplier or a photo-multiplier and scintillation detector. Results from these machines appeared originally as numbers from a printer and this information was punched onto cards, by hand, for feeding the computer. Punched paper tape is now being substituted and this can be fed straight into the computer. The scalers in use now have a dead time of about 60 ns. We have used faster apparatus[2], down to about 15 ns or less, and this will be introduced for 'customer' sample analysis as soon as we are confident that it will be entirely reliable when in everyday use. Again, the programme ISOTOP is used, with various sub-programmes to suit special cases.

A single stage of one machine is sometimes used for light elements and we have arranged to switch the magnetic field using n.m.r. control. Fields in the region of 800 G have been switched some 10% with a delay time of about 2 s and an accuracy of 30 p.p.m.

The data from our AEI MS5 machines is now mostly handled using a digital voltmeter and paper tape punch. The mass spectrometer itself is not much altered, save to take an output from the electrometer amplifier which feeds the chart recorder, and to fix a small magnet to the 'magnet current control' shaft which operates a pair of reed switches. This arrangement gives an indication when the scanning has reached the end of its sweep. A digital voltmeter reads the output from the spectrometer and is arranged to sample at about three times each second. This output is punched onto paper tape. The operator sets up the machine so that the required spectrum is obtained, within the range of the automatic scanning control, and sees that the largest peak does not overload the amplifier. He then switches on the automatic punching apparatus. We find that scanning can be much faster than is normally used for hand-reading the recorder chart. This raw data then goes straight to the computer, where the programme DRAMS is used. This works in the following way. A number of runs on a single paper tape are used to set up a file of data on magnetic tape. Any of these files can be selected for further processing, as required.

The raw data is checked for punch errors, interpolating for points which are wrong and regrouping into a format suitable for the rest of the programme. The distances, in terms of the number of points, between the sweep end markers on the tape are used to select sections of data for consistency of scanning rate and the limits on the mass range being scanned. If necessary, a run is automatically broken down into sections which satisfy these conditions and any remaining data for this run is ignored.

The rest of the programme can be split into two main parts—a pattern recognition type for selecting the peaks for ratio measurement and a programme for producing ratios with appropriate statistics for output to the customer.

This section is basically the same as the ISOTOP programme used on the two-stage machines with peak switching.

(*i*) The pattern recognition process relies on the fact that since all the masses are being scanned a number of times, the distance between any given mass and its next appearance should be similar to the distance between the markers.

The data is first searched for peak and trough positions so that this will include all real peaks and possible sparks or 'noise peaks'. The peak positions are scanned to find sets which occur at similar intervals and their positions relative to the nearest low mass end marker are used to build up a dummy set of positions within a scan.

This dummy set of positions is compared with a list of expected mass values for this sample and hence mass numbers are allocated to the dummy set of positions.

All the peaks in the original data are then compared against the dummy set to decide whether they fall into the categories of real peaks, misplaced real peaks, spark or noise, or missing peaks (i.e. a dummy set position for which there is no corresponding peak).

(*ii*) The section of the programme which gives ratios, then uses sets which have real or misplaced real peaks corresponding to each dummy set position. A part of the programme can be used to write the peaks on a graph plotter from the raw input data so that an examination of the form of the spectrum can be made if required for checking purposes. We have had some trouble with this programme. The main problem has been in recognizing peaks which are similar in magnitude to the zero level. Large peaks can easily be detected by using a discriminator and also measuring the width of a peak, but for the small peaks it is difficult for an operator to set a discriminator level accurately and consistently. After trying various types of discriminator function, the system which I have just described, has been used. This system does not rely so heavily on the operator and, in consequence, is easier to use.

In some cases it is also difficult for the programme to allocate the correct mass marker to each position in the dummy set. In this case it is sufficient to allocate mass numbers to the positions after a run.

Results show that the programme will select the correct peaks 95% of the time, when there is up to ± 5% variation in the position of the end markers or in the position of individual peaks within a complete run.

The worst errors in peak measurement are for the very small peaks. For a set with 12 db signal-to-noise ratio the average error for a typical run is ± 3·5% with a spread of ± 6%.

As an additional check, the final values used in the ratio section only include peaks that give values which fall within a normal distribution.

Further work on the specification of base and peak values is being carried out to reduce the spread on these measurements. We are considering the installation, in the department, of an on-line computer which would take up all our data from electrical interfaces. This would give a number of advantages, including immediate assessment of results by the operator, quicker turn round, and the possibility of more sophisticated machine control.

I would like to describe briefly some of the new instrumentation we have developed for our mass spectrometers. We are faced with the problem of servicing and keeping running some 12 machines of one sort or another which originally had a frightening diversity of electronic equipment. It is clear that there are many functions which are common to several machines, and we have now embarked on a programme to buy, adapt, develop or design suitable common units. The guiding principles are that they should be as self-contained as possible, since this makes for quick replacement and easier servicing, that solid-state circuits should be used, if at all possible, since these are more reliable and have less fire and shock risk, and that a 19 in. unit size (which may contain replaceable sub-units) should be used since commercial boxes are mostly available in this form, and we use them wherever possible. So far we have had the following units manufactured and tested for some months. In all cases quantities of at least six and sometimes many more have either been made or are planned as soon as the prototypes have been thoroughly evaluated.

(*i*) An emission stabilizer which is an all solid-state box, made to work at up to 10 kV from ground, and runs a gas source filament regulating emission to 1 part in 10^4 per hour over a range of 10 μA to 2 mA. The electron voltage is variable from 50 V to 200 V and the unit is proof against e.h.t. sparks.

(*ii*) A magnet sweep unit which has six ranges with scan times of $7\frac{1}{2}$–240 min. Any of these can sweep the whole or a portion (down to 1/30) of the magnet current. The scan can be made to start at any point and will return to this point when reset. There is provision for external programming, and this gives an output whose rate and direction is proportional to the input voltage, with a maximum of 120 V giving the fastest scan.

(*iii*) A magnet supply unit, designed to our specification by industry, which feeds an 8 Ω coil with a maximum of 5 A. This gives a mass range up to about 400 with one of our 12 in. machines. Stability is better than 1 in 10^4 per hour and the sweep unit which I have just mentioned fits into this box.

(*iv*) A comparator unit, which accepts inputs from the acceleration supply or magnet and feeds a suitable proportion out to a digital voltmeter. This allows us to watch the stability of these or any other parts of the electrical supplies, and to make measurements of beam width, etc.

(*v*) Voltage divider units to supply the ion source potentials. This has involved special switches to handle the high voltages.

(*vi*) Although I think we will now use commercial, high voltage supplies for both photo-multiplier and acceleration potentials, we have designed and built some six, 6 kV units with a stability of 1 part in 10^4 per hour which are 'short-circuit proof' and recover quickly to their set voltage. These use all semiconductors except for one high voltage shunt stabilizer valve.

(*vii*) A filament stabilizer for a solids machine. This feeds a side and main filament with d.c., stabilized to 1 part in 10^4 against load change, and has a variable and 'stiff' bias supply built in. It can be switched to a simple electron beam regulator, if required for use with an electron bombardment source[3].

(*viii*) A timer, which programmes the switching from one mass position to another, and allows us to select any combination of up to eight positions, for any selected time between $\frac{1}{2}$ s and 64 s. There is also an arrangement which allows the cycle to be repeated '$4n + 1$' times and then stops, so that the operator can have a look at that set of results.

REFERENCES

1 RIDLEY, R. G., and YOUNG, W. A. P. *Instrum. Pract.* 20 (1966) 321

2 YOUNG, W. A. P., RIDLEY, R. G. and DALY, N. R. *Nucl. Instrum. Meth.* 51 (1967)

3 TYRRELL, A. C., RIDLEY, R. G. and ROBERTS, J. W. *J. scient. Instrum.* 42 (1965) 806

SYMPOSIUM IMPRESSIONS

Professor J. C. Robb

University of Birmingham

PROFESSOR Robb in summing up his impressions of the meeting felt that although the number of scientists in the field of mass spectrometry far exceeded that of 20 years ago, the essential cameraderie still existed and was encouraged by meetings such as this.

In tracing the history of mass spectrometry since the time of J. J. Thomson, Professor Robb related its development to the improvement in photographic techniques, which initially had imposed a severe restriction on the work of Aston, and to the advances made in the field of applied electronics. Thus, since the early 1930s, stabilized d.c. supplies, stabilized amplifiers, sensitive detectors and recorders have become increasingly available and many of the difficulties that confronted Dempster, when he first devised electrical detection techniques, no longer exist.

Interest in petroleum fractions, during World War II, accelerated the development of mass spectrometers but, in this country, it was after the war that the most remarkable advances in instruments and techniques were witnessed. Early spectrometers had resolutions (if one was lucky) of 220 and the tedious but essential process of carbonizing source filaments, in order to produce stable emission characteristics, was recalled by Professor Robb with some nostalgia. In similar vein, the speaker referred to the early desire to get rid of metastable peaks which were regarded as an embarrassment but which nowadays can be utilized to provide essential structural information.

Professor Robb speculated on the way in which vapour phase chromatography had developed after mass spectrometry and wondered how the latter would stand today if the reverse had been the case. As it was, the advent of gas chromatography had relegated mass spectrometry for a time to the realm of pure research until the early 60s when the separatory powers of gas chromatography were first combined with the analytical powers of mass spectrometry. Now, of course, the on-line combination of the two techniques is one of the most powerful tools available to the chemist.

Of the 21 papers delivered, Professor Robb noted that nine were from the manufacturing industries. He felt that this was a good feature of the symposium and complimented the speakers on their restraint in not transgressing into the field of advertizing. Five papers dealt with the cracking or decomposition of ions and he thought that here the feedback of information to the organic chemist in particular, was a firm step in the right direction. Three papers were concerned with surface or catalytic interests, a field which was just opening to the mass spectroscopist, and the results quoted showed considerable promise. One paper on flavours research showed an interesting application of the gas chromatography–mass spectrometry combination which, in fact, was the only way available for this type of research problem.

In general, it was felt that the future would see rapid progress in the development of the quadrupole mass filter and finally, in data reduction by computer link-up.

INDEX

www.ingramcontent.com/pod-product-compliance
Lightning Source LLC
LaVergne TN
LVHW102103300926
849161LV00001B/6
* 9 7 8 0 8 2 0 6 0 0 1 6 1 *